GUIDE TO
ECOLOGY
INFORMATION AND ORGANIZATIONS

GUIDE TO

ECOLOGY

INFORMATION
AND
ORGANIZATIONS

———◆•■•◆———

JOHN GORDON BURKE

JILL SWANSON REDDIG

THE H. W. WILSON COMPANY

NEW YORK

1976

Library of Congress Cataloging in Publication Data

Burke, John Gordon, 1938–
   Guide to ecology information and organizations.

   Includes index.
   1. Human   ecology—Bibliography.   2. Human ecology—Directories.   3. Ecology—Bibliography.   4. Ecology—Directories.   5. Environmental protection—Bibliography.   6. Environmental
protection—Directories.   I. Reddig, Jill Swanson,
joint author.   II. Title.
Z5861.B87 [GF41]   016.30131   75-45400
ISBN 0-8242-0567-7

TABLE OF CONTENTS

ACKNOWLEDGMENTS

We wish to acknowledge with appreciation the help of individuals and organizations whose cooperation enabled us to complete this book. We are particularly grateful to the conscientious respondents who made our task of compiling this information easier. In addition, we would also like to acknowledge the extensive use made of the resources of the following libraries: Boston Public Library; Chicago Public Library; Evanston, Illinois, Public Library; Northwestern University Libraries, in Evanston, Illinois; Oak Park, Illinois, Public Library; Rosary College Library, in River Forest, Illinois; San Antonio, Texas, Public Library; Skokie, Illinois, Public Library; Trinity University Libraries, in San Antonio, Texas; and the University of Kansas Libraries, in Lawrence, Kansas. Special thanks are due to Paula Davis. We wish to dedicate this book to Robert Reddig for his patience and encouragement.

INTRODUCTION

Ecology has come to be defined as the interrelation between living organisms and their environment; the roots of this concept have been traced by historians of science to Aristotle. While there is a clearly identifiable history of scientific inquiry along these lines beginning with Aristotle and continuing into the nineteenth century, it was not until 1869 that the term *œkologie* or *ecology* was coined by Ernest Heinrich Haeckel, a German biologist and philosopher. By the beginning of the twentieth century, ecology was an established field, and there was considerable debate as to whether Haeckel's *ecology* or some other term should be used to designate the new science.

Ecology, in the natural sciences, found formal expression in animal ecology, plant ecology, hydrobiology (oceanography, limnology), wildlife management, and bioclimatology or biometeorology. In the social sciences, ecology evolved as an operative concept in sociology, geography, population studies, and a number of other disciplines. Concern over the pollution of the environment developed rapidly after the publication of Rachel Carson's *Silent Spring* in 1962, and on April 22, 1970, the first Earth Day, the issue exploded into the national media and the popular consciousness. Earth Day, marked by demonstrations and educational programs, was a nationwide event organized by a Washington group, Environmental Teach-In, subsequently Environmental Action, Inc. Although for many years various organizations and individuals had been calling attention to environmental problems, it was this event which signalled a revolution in public sentiment.

After Earth Day, the role human beings play in violating the environment became a major topic of concern. The problems of identifying, controlling, and repairing damage to the environment are central to the concept of ecology as it is used in this book. To provide background on the development of this concern, we have included a number of information sources that were written within subject disciplines from the standpoint of ecological principles. Most of our listings, however, are interdisciplinary in nature.

Our purpose is to provide some structure to the proliferating literature of the field. The most important element in this book is the index, which identifies material in specific subject areas. We hope that we have located sources for most of the information demanded today by public librarians and public library patrons.

We have written this book primarily for public librarians, but hope that it will also be helpful to librarians in other settings and to library patrons. It has been our intent to provide a general introduction to the various categories of information sources, so that the average library patron will be able to use this book without difficulty.

Since we are selecting materials appropriate for public libraries, we have excluded most technical sources. However, we have discovered that rather technical material is essential in two types of cases: (1) subject areas, such as quantitative ecology, which are technical by nature, and (2) specific problems, whose analysis and solution may require technical data. While environmental problems are usually clearly identified in popular literature, solutions are not. The complexity of the solutions to difficult environmental problems cannot be avoided. Therefore, we have included technical material where necessary—for example, on the economics of environmental quality control.

In our search for information and resources, we have not discriminated against any format. The reader will find here both print and nonprint materials, as well as the names of persons who are willing to share their expertise.

With the wide dissemination of educational materials on the environment and its problems, it is easy to be optimistic about finding adequate solutions to the environmental crisis. To us, however, it seems clear that the effective resolution of many environmental problems will come only when citizens demand that these problems be solved and effectively mount political opposition to those who would frustrate their efforts. We hope this book will be useful in that never-ending struggle.

CHAPTER

1

CITIZEN ACTION GUIDES

———————————◄•●•►———————————

Ecology action guides, now indelibly part of the popular bibliographical consciousness, owe their existence to the 1970 Earth Day. This event produced the *Earth Tool Kit: A Field Manual for Citizen Activists* and established a formula for citizen action on environmental issues.

This book has generated many similar titles, with the Sierra Club issuing a number of key books on a variety of subjects. It is the *Earth Tool Kit*, however, which remains the prototype, to which all other titles must, in one way or another, be compared.

1-1 *Earth Tool Kit: A Field Manual for Citizen Activists.* Prepared by Environmental Action, edited by Sam Love. New York: Pocket, 1971. 360p.

The first ecology action guide, complete with guidelines for personal and community ecological inventories, a list of ecology action groups, and a glossary.

1-2 *Ecotactics: The Sierra Club Handbook for Environment Activists.* Edited by John G. Mitchell with Constance L. Stallings. New York: Pocket, 1970. 288p.

This guide contains case histories of environmental action as well as a bibliography entitled "The Activists' Bookshelf" and a directory of the Sierra Club's chapters and other local groups.

1-3 *The Do-It-Yourself Environmental Handbook.* Rev. ed. Prepared by the Dayton Museum of Natural History, Dayton, Ohio, and edited by E. J. Koestner, Joseph J. McHugh, and Ralf Kircher. Boston: Little, Brown, 1972. 76p.

A checklist for personal action, centered around the home, consumer purchasing, and so forth.

1

1-4 *Everyman's Guide to Ecological Living.* By Greg M. Cailliet, Paulette Y. Setzer, and Milton S. Love. New York: Macmillan, 1971. 119p.

Arranged by resource to be conserved. Specific action is recommended for each category. Contains a list of recommended information sources and concerned organizations.

1-5 *The Ecological Citizen: Good Earthkeeping in America.* By Dirck Van Sickle. New York: Harper, 1971. 295p.

An entertaining account of environmental action. Abandons the traditional "practical" format for a more literary, if not so straightforward, presentation of tactics and information.

1-6 *Man's Impact on the Global Environment: Assessment and Recommendations for Action.* Report of the Study of Critical Environmental Problems (SCEP), sponsored by the Massachusetts Institute of Technology. Cambridge: MIT Press, 1970. 319p.

Concerned with worldwide environmental problems, this report contains extensive recommendations. A bibliography accompanies each work-group's paper.

1-7 *Action for Wilderness.* Edited by Elizabeth R. Gillette. New York: Sierra Club, 1972. 222p.

Based on the Sierra Club's twelfth biennial wilderness conference, this was produced as a "guide for waging wilderness campaigns." Appendixes contain the text of Public Law 88-577, the Wilderness Act, and a list of the present and potential units of the National Wilderness Preservation System.

1-8 *Space for Survival: Blocking the Bulldozer in Urban America.* Edited by Charles E. Little and John G. Mitchell. New York: Pocket, 1971. 228p.

A guide for land-saving efforts in the city. Through case histories, it presents techniques for saving land for human access in urban areas.

1-9 *Blueprint for Survival.* By the editors of *The Ecologist*: Edward Goldsmith, Robert Allen, Michael Allaby, John Davoll, and Sam Lawrence. Boston: Houghton Mifflin, 1972. 189p.

This title has been reorganized internally for U.S. publication. It offers a strategy for change to achieve a "stable society." In separate chapters, it deals with social systems, population and food supply, and nonrenewable resources.

1-10 *The Environmental Handbook.* Edited by Garrett De Bell. New York: Ballantine, 1970. 365p.

Prepared for the First National Environmental Teach-In. Contains sections on meanings of *ecology* and *ecotactics*. A bibliography and film list are appended to this collection.

1-11 *The Voter's Guide to Environmental Politics Before, During, and After the Election.* Edited by Garrett De Bell. New York: Ballantine, 1970. 305p.

An attempt to educate the voter by identifying ecologically-minded politicians. Outlines techniques for making ecological issues into political issues.

1-12 *The User's Guide to the Protection of the Environment.* By Paul Swatek. New York: Friends of the Earth/Ballantine, 1970. 312p.

An action guide that stresses using the consumer's purchasing power for effective ecological change. Chapters on shelter, gardening, water use, energy use in the home, cosmetics, clothing, transportation, and recreation, as well as a bibliography of periodicals of interest to the environmentally-concerned consumer.

1-13 *The Population Activist's Handbook.* Population Institute. New York: Macmillan, 1974. 176p.

This is a practical and well-planned guide to action. It tells the reader how to organize, get publicity, effect legislation, and educate the community. While it has an excellent section on policy, this book concentrates on action, with specific suggestions for setting up and promoting educational programs, professional health services, programs to promote contraception, and so forth. One of the five sections of the book is devoted to action programs for high-school students, and a special chapter is included for college students on establishing population-related courses and services. Information sources are provided throughout the book, including not only bibliographies but also names and addresses of relevant organizations, libraries with population collections, and colleges with courses in human sexuality. Sources for both print and nonprint materials are listed as well. Indexed.

1-14 *The Grass Roots Primer.* James Robertson and John Lewallen. San Francisco: Sierra Club, 1975. 287p.

An up-to-date and practical action guide. It features eighteen documented programs of local environmental activists in the United States and Canada. There is a chapter on planning, organizing, and carrying out a local environmental project and a chapter on how local activity can make federal and state environmental legislation work for environmental improvement. Included also are a bibliography of information sources about the environment and a directory of current environmental projects in the United States and Canada. Because of its practical approach and its current information, it is probably the most useful of the action guides now available.

CHAPTER

INDEXES AND ABSTRACTS

The term *ecology* was first used as a separate subject entry in the *Readers' Guide to Periodical Literature*, the standard general index to periodicals, in the 1922–1924 cumulative volume. The heading continues, of course, to be used today. A great deal of material on ecology has been indexed by *RG* in the past decade, and there has also been an increasing amount of material listed in *Social Sciences Index* and its predecessor, *Social Sciences & Humanities Index*, as well as in *Biological & Agricultural Index*. The researcher in need of general information about ecology, however, will probably find the *Public Affairs Information Service Bulletin* the single most helpful general index.

2-1 *Public Affairs Information Service Bulletin*, 1915–   . New York: Public Affairs Information Service, Inc.

A selective index to materials in the fields of economics and public affairs. Pamphlets, current books, periodical articles, and government documents are included, as well as "any other useful library material." More than 1,000 periodicals are selectively indexed each year. The *Bulletin* is published weekly, except for the last two weeks in each quarter, and cumulated five times a year. The final cumulation is the annual volume. For works in French, German, Italian, Portuguese, and Spanish, a cumulated volume entitled *Public Affairs Information Service Foreign Language Index, 1968–1971* was published in 1972. *PAISFLI* is now issued quarterly, and the fourth quarterly issue is the annual cumulation. Both indexes should be consulted under one or more of the following subject headings: Ecology, Environment, Human ecology, Man—influence of environment, and Conservation of resources.

In recent years, several new indexes have begun publication.

2-2 *Environmental Periodicals Bibliography: Indexed Article Titles*, 1972–   . Santa Barbara, CA: Environmental Studies Institute.

Indexes feature and review articles on the environment from approximately 260 U.S. and foreign periodicals. Published six times a year; a seventh issue provides the annual index. Each issue is divided into six subject category listings: General, Human ecology, Air, Energy, Land resources, Water resources, and Nutrition and health. *Environmental Periodicals Bibliography* reproduces the table of contents of each magazine indexed, thus providing author and subject access to the material. Each issue also carries a list of periodicals indexed, with pages cited for the location of their tables of contents within individual issues.

2-3 *Environment Information Access*, Jan. 29, 1971–   . New York: Environment Information Center, Inc.

A semimonthly journal with abstracts of ecology material appearing under twenty-one categories. Citation includes references to charts, graphs, and photos, where present. In addition to abstracts of newspapers and periodical articles, documents and technical reports, *Access* carries, in separate sections, reviews of environmental books, an index and abstract of all environmental entries in the *Federal Register*, a conference calendar, and an "issue alert," a guide to significant information in each journal. A document accession number index, subject index, industry index, and author index are also included in each issue.

2-4 *The Environment Index,* 1971–   . New York: Environment Information Center, Inc.

The annual cumulated index to *Environment Information Access* (above) and a broad-spectrum index to materials on ecology. In addition to indexing periodicals, newspapers, documents, and providing listings of books and selected films, this cumulation provides an analysis of the year's significant events affecting the environment, a status report and synopsis of current federal environmental legislation, a roster of the year's major meetings, conferences and symposia, a directory of state environmental control officials, and a list of environmental patents issued by the U.S. Patent Office.

Indexing of material is provided by three separate sections arranged by subject, industry, and author. The subject listing includes organizational as well as geographic entries, while the industry index is arranged by major industries according to Standard Industrial Classification number (SIC). Entries in subject and industry indexes do not identify authorship. The author index refers to accession numbers, which dictates that *Environment Information Access* must be used, by means of an accession number index, in connection with *The Environment Index* for searches when only the author's name is known.

The material indexed in *The Environment Index*, with the exception of monographs and films, is available from the publishers on microfiche or hard copy, individually by subject category or on a full subscription arrangement.

2-5 *The Environment Film Review*, 1971–   . New York: Environment Information Center, Inc.

An annual guide to environmental films, with a total of 627 reviewed and annotated in the first edition. Entries include running time, purchase price and rental fee, release date, sponsor where given, producer and/or distributor, and intended audience. All films are annotated, and a rating system is employed. One star is given to "above average" films, and two stars, to "exceptional" films.

An index section contains an alphabetical index, subject index, industry index, and sponsor index to films. Cross references are made from one film to another "when appropriate," and a list of addresses of film distributors precedes the review section.

2-6 *Pollution Abstracts*, 1970–   . San Diego, CA: Pollution Abstracts.

A bimonthly publication covering the entire spectrum of pollution control and research. It abstracts and provides key-word and author indexing for technical and nontechnical information in this area. Air pollution, water pollution, land pollution, and general pollution are the nontechnical topics covered. For technical material, the index covers air pollution, fresh water pollution, marine pollution, land pollution, noise pollution, sewage and waste treatment, general pollution, contracts, and patents. A Calendar of Meetings concerned with pollution is featured, as are occasional short articles about various aspects of environmental research. A document retrieval service for items indexed is available upon application.

CHAPTER

3

REFERENCE BOOKS

With so little attention, until recently, devoted to ecology as a separate subject, the classic reference book is more often than not written in another discipline, such as biology or geology. We have attempted to list books which provide a bibliographical foundation for reference information in ecology, as well as many of the generally excellent reference books written since ecology emerged as a discipline in its own right.

Reference books have been categorized into the following types: bibliographies; dictionaries; directories; encyclopedias; handbooks, guides and manuals; and others. The latter category is used to accommodate several titles which do not fit elsewhere, but which we felt it necessary to include.

BIBLIOGRAPHIES

3-a-1 Dee, Sandra R. *A Basic Environmental Collection*. Monticello, IL: Council of Planning Librarians, 1973. (Exchange bibliography, 410) 15p.

> A bibliography of seventy-four basic items for an "environmental collection." It lists films, records, games, plays, bibliographies, periodicals, and monograph material. Items are annotated and are divided into a general book section and a special materials section.

3-a-2 Durrenberger, Robert W. *Environment and Man: A Bibliography*. Palo Alto, CA: National Press Books, 1970. 118p.

> There are 2,225 entries in this bibliography, representing, for the most part, material published since 1960. Special emphasis,

according to the compiler, was placed upon "the relationships of man and the environment," "the arid lands of the world," and the "problems associated with the atmosphere and the hydrosphere." A topical subject index identifies citations by forty-six subject categories, and a brief appendix on federal legislation is included.

3-a-3 Ehler, Charles N. *Environmental Systems Planning and Management: A Preliminary Sorting of Literature.* Monticello, IL: Council of Planning Librarians, 1972. (Exchange bibliography, 251) 64p.

Designed to "familiarize the urban planner with . . . environmental systems planning and management." This bibliography is organized into seven different subject areas: the environment, ecology, population, general systems theory/cybernetics, environmental technology, technological innovation, and technology assessment. Items are listed alphabetically within each category and are not annotated.

3-a-4 *Index to Ecology.* 2d ed. Los Angeles: National Information Center for Educational Media, 1973. 212p.

A bibliography of multimedia materials divided into a subject guide, an alphabetical listing of entries, and a directory of producers and distributors. Entries include title with subtitle, size and physical description, length, stock or color code, description of contents, series title when applicable, audience level, producer and distributor code, year of release, LC number when available, and broadcast quality for video tapes. Materials listed include 35mm filmstrips, 8mm motion picture cartridges, 16mm motion pictures, video tapes, records, audio tapes, and transparencies.

3-a-5 Knobbe, Mary L. *Air Pollution: A Non-Technical Bibliography.* Monticello, IL: Council of Planning Librarians, 1969. (Exchange bibliography, 83) 9p.

An annotated listing source designed to meet the information needs of local officials desiring nontechnical information on air pollution. Items are grouped in the following categories: air pollution studies by city, county or state; control devices and methods; legislation-administration-economics; source and source-emission studies; and standards and criteria. Most items listed are articles published in periodicals.

3-a-6 Meshenberg, M. J. *Environmental Planning: A Selected Annotated Bibliography*. Chicago: American Society for Planning Officials, 1970. 79p.

Material listed ranges from works on general environmental issues to monographs on flood plain management and landscape analysis. The practical application of "planning concepts and techniques" is emphasized. Four hundred and seventy-three publications are listed. Annotated.

3-a-7 Minneapolis. Public Library. Environmental Conservation Library. *Book Catalog of the Environmental Conservation Library, Minneapolis Public Library*. Chicago: American Library Association, 1974. 201p.

Approximately 2,800 titles in the library's collection are included in an author and main-entry catalog listing. Subject and title catalogs are included, in addition to a list of subject headings for vertical files and a list of recommended periodicals.

3-a-8 Sangster, Robert Powell. *Ecology: A Selected Bibliography*. Monticello, IL: Council of Planning Librarians, 1971. (Exchange bibliography, 170) 27p.

A bibliography which emphasizes more recent publications and provides listings under the following topics: general, biogeochemical cycles, communities, ecological mentality, ecosystems—general and specific, energy flow, and populations. Items that are considered "superior" by the compiler are annotated.

3-a-9 University Microfilms. *A Bibliography of Doctoral Research on Ecology and the Environment*. Ann Arbor, MI: University Microfilms, 1973. 92p.

A guide to those dissertations accepted at American universities from 1938 to 1970 that are available from University Microfilms. The guide lists more than 900 dissertations on ecology and the environment. It was produced by a retrospective search of *Dissertation Abstracts International*. Arranged by author, with full title of dissertation, name of accepting institution, and date of completion.

3-a-10 Van Der Leeden, Fritz. *Ground Water: A Selected Bibliography*. Port Washington, NY: Water Information Center, 1971. 116p.

Approximately 1,500 references under thirty-two topics, each dealing with a particular aspect of the field of ground-water hydrology. English-language material predominates, but foreign-language material is also included. Materials are listed by type (books, journals and bulletins, general bibliographies) as well as by topic (well-logging, land subsidence, and even water-witching).

3-a-11 Watkins, Jessie B. *Ecology & Environmental Quality: A Selected and Annotated Bibliography for Biologists and Earth Scientists.* Syracuse, NY: Syracuse University Libraries, 1971. 127p.

A guide to the Syracuse University Library collections in ecology. Annotated.

3-a-12 Woodrow Wilson Memorial. *The Human Environment.* Washington, DC: Woodrow Wilson International Center for Scholars, 1972. (Environment series 201) 2 vols. 171p., 109p.

Volume 1: subtitle, *A selective, annotated bibliography of reports and documents on international environmental problems.* Includes official documents that were prepared for the United Nations Conference on the Human Environment, Stockholm, June 1972, and documents (including doctoral dissertations) prepared by or for participating governments and organizations attending other international meetings. The documents emphasize "the political, legal, economic, social and institutional aspects of environmental problems of global, national or more than local significance." Arranged by issuing agency and/or author. Includes subject index, author index, and a numerical list of United Nations Documents.
Volume 2: subtitle, *Summaries of national reports on environmental problems.* This work contains over seventy summaries of reports, arranged alphabetically by country.

3-a-13 Winton, Harry N. M. *Man and the Environment: A Bibliography of Selected Publications of the United Nations System, 1946–1971.* New York: Unipub/Bowker, 1972. 305p.

Entries are annotated and are grouped under a number of specific subject headings—Seismology, for example. Sales numbers and document symbols are provided for all entries. Indexed by author, series, title, and subject.

DICTIONARIES

3-b-1 Carpenter, John Richard. *An Ecological Glossary*. Norman: University of Oklahoma Press, 1938; reprinted by Hafner, 1971. 306p.

A standard dictionary of terms and concepts employed in the fields of botany and zoology. A useful introduction records the history of the development of scientific nomenclature. A bibliography of literature cited in the glossary is included as well as a historical bibliography arranged by date of publication. Because of its publication date, readers seeking a definition of terms in current use—*ecosystem*, for instance—will have to look elsewhere.

3-b-2 Sarnoff, Paul. *The New York Times Encyclopedic Dictionary of the Environment*. New York: Quadrangle, 1971. 352p.

A dictionary of technical environmental terms, arranged alphabetically. Abbreviations and acronyms are included. A number of illustrations appear throughout the work.

DIRECTORIES

3-c-1 Brown, Patricia L. *Energy Information Resources: An Inventory of Energy Research and Development Information Resources in the Continental United States, Hawaii and Alaska*. Washington, DC: American Society for Information Science, 1975. 207p.

This directory lists almost 200 federal and state agencies, professional and trade associations, and corporations, in addition to more than 100 publications and publishers of information about energy. The volume is indexed by organization, acronym, geographic location, and subject.

3-c-2 *Conservation Directory*. Washington, DC: National Wildlife Federation. Annual.

A directory of organizations concerned with conservation. Major national and state organizations are included; both public and private organizations are listed.

3-c-3 *Directory of Consumer Protection and Environmental Agencies*. Edited by Thaddeus C. Trzyna and the staff of the Center for California Public Affairs. Orange, NJ: Academic Media, 1973. 627p.

The section on environmental agencies makes up the larger portion of this work. Both governmental and private agencies are listed, at state, national, and international levels. Addresses, telephone numbers, key personnel, and descriptions of purpose and activity are given for most agencies. Indexed by publication, personnel, organization, and subject area of organizational concern.

3-c-4 *Directory of Environmental Information Sources.* 2d ed. Edited by Charles E. Thibeau. Boston: The National Foundation for Environmental Control, 1972. 457p.

Listings in this directory are organized by information sources, such as government agencies, citizens' organizations, and trade associations; abstracts, directories, and indexes; published bibliographies; conference and symposium proceedings; documents and reports; serials; and books, films, and filmstrips.

Selection criteria are outlined before each chapter. Entries for government agencies generally include address, telephone number, and summary of responsibilities, while entries for citizens' organizations include address, telephone number, names of key officers, and publications. Both the book and film lists are organized by subject, but the film titles are annotated while the book titles are not. Appendixes include a list of acronyms with the full name and address of each organization, a key to geographic location of citizens' organizations, and a topical index.

3-c-5 *Directory of Government Agencies Safeguarding Consumers and Environment.* 5th ed. Alexandria, VA: Serina, 1973. 135p.

A monograph organized by area of concern, such as food and drugs, meat and poultry surveillance, consumer protection, air-pollution control, and noise abatement. Within these subject areas, agencies are listed by state, with the appropriate federal jurisdictions noted first, then state and local agencies. Names, addresses, and telephone numbers are provided. Not indexed.

3-c-6 *Directory of Organizations Concerned With Environmental Research.* Compiled by Wendell A. Mordy and Phyllis A. Sholtys. Fredonia, NY: Lake Erie Environmental Studies, State University College, 1970. 150p.

An international directory keyed by field of interest, type of agency, and size of staff. (See 3-c-13 for revised edition.)

3-c-7 *The Energy Directory, 1974–   .* New York: Environment Information Center, Inc. Annual. 1974 ed. 418p.

A guide to government agencies, at the state, regional, national, and international levels, and to private organizations and industries concerned with energy. Gives the names, addresses, and telephone numbers of organizations and agencies, and the names of many key personnel. Some information on activities and services. Indexed by organization, subject, location, and SIC number.

3-c-8 National Audubon Society. Nature Center Planning Division. *Directory of Nature Centers and Related Environmental Educational Facilities.* New York: National Audubon Society, 1971. 72p.

This directory identifies facilities in the United States and Canada which are actively involved in environmental education. It contains 459 entries, including 36 for Canada. Facilities are listed alphabetically by state or province and community. Individual entries contain the name, address, and telephone number of each facility, and the organization responsible for its operation and maintenance. Other information noted: size of facility, type of facility when not implied in its name, formal programs offered, special features, the availability of guided or unguided tours, and the hours the facility is open.

3-c-9 Noyes Data Corporation. *Pollution Control Companies U.S.A.* Park Ridge, NJ: Noyes, 1972. 239p.

Intended as a directory to all industries involved in the field of environmental control. The first section lists approximately 1,500 companies that manufacture or supply products in this area. The second section lists more than 500 companies that provide professional services. When available, number of employees and annual sales are given.

3-c-10 Onyx Group, Inc. *Environment U.S.A.: A Guide to Agencies, People, and Resources.* New York: Bowker, 1974. 451p.

A directory which lists government environmental agencies (federal and state), private organizations, environmental consultants, environmental officers of American corporations, environmental libraries, films, and books. Chapters have also been prepared on labor unions and the environment, environmental employment, environmental educational programs, environmental fund raising, and the environment and the law. Meetings and conferences, classified by topic, are also listed. The book also contains an extensive bibliography, a glossary, and an index.

3-c-11 Smithsonian Institution. *National and International Environmental Monitoring Activities: A Directory.* Washington, DC: Smithsonian Institution Press, 1970. 292p.

Basic information on the number and distribution of present and planned environmental monitoring programs.

3-c-12 *Who's Who in Ecology 1973.* New York: Special Reports, 1973. 291p.

A directory of individuals who are involved in one way or another with ecological concerns. The following information is supplied for each entry: name, title, place of birth, father's name, mother's name, education, marital status, children, career summary, professional activities, civic activities, nonbusiness directorships, honors, lodges, fellowships, and publications.

3-c-13 Wilson, William K. *World Directory of Environmental Research Centers.* New York: Bowker, 1974. 330p.

A revised edition of the *Directory of Organizations Concerned With Environmental Research*, compiled by W. A. Mordy and P. A. Sholtys (see 3-c-6). Entries include the names and addresses of research organizations in the following subject areas: biology, chemistry, engineering, geology, physics, social science/humanities, and interdisciplinary areas. Some of the entries are annotated. Arranged by subject and geographical location.

3-c-14 Wolff, Garwood R. *Environmental Information Sources Handbook.* New York: Simon & Schuster, 1974. 568p.

This directory is intended to provide an "in-depth . . . bibliographic inventory of the major environmental information resources in the United States." Listed are national, civic, and conservation groups, engineering and professional societies, industry associations, commercial newsletters, offices and agencies of the federal government, regional organizations, state and interstate agencies, university and private study centers, and trade and business magazines. Indexing access is provided for abstracting, indexing, and bibliographic journals, curriculum guides, directories, film lists, libraries, periodicals, and published abstracts, indexes, and bibliographies. While entries are rather wide-ranging in scope, this is a useful directory of information sources about ecology.

3-c-15 *World Directory of Environmental Education Programs.* Edited by Philip W. Quigg. New York: Bowker, 1973. 289p.

A guide to postsecondary environmental training and study in seventy countries. Entries include name and address of institution, description of program, degrees conferred, and tuition information. Arrangement is by country. The book has a subject index by field of specialization and a series of appendixes which further define the types of training available at various institutions.

3-c-16 *World Directory of Environmental Organizations.* Preliminary ed. Edited by Thaddeus C. Trzyna. Claremont, CA: Sierra Club, 1973. 155p.

Published in English and French, the directory is organized in three parts. Part 1 covers international organizations and is divided into sections on intergovernmental agencies, the International Union for Conservation of Nature and Natural Resources (IUCN), and nongovernmental organizations. Part 2 lists national and selected regional and local organizations in each of the world's nations. Basic geographic, demographic, and financial data are provided for each country as well as descriptions of national directories. Part 3 is a listing of organizations concerned with the environments of territories, such as New Guinea and the Galápagos Islands, which are important as unique natural communities.

Each organization entry includes the name and address of the organization with an English translation provided when possible. If available, the following information is also included: names and titles of key personnel, periodical publications, important affiliations, and the major interests of the organization, indicated by an abbreviation. Indexes to English titles of international organizations listed in Part 1 and acronyms of those organizations complete the volume.

ENCYCLOPEDIAS

3-d-1 Fairbridge, Rhodes Whitmore. *Encyclopedia of Atmosphere Science and Astro-geology.* New York: Reinhold, 1968. 1,200p.

This is the second volume of an eight-volume series entitled Encyclopedia of Earth Sciences. The editor, Rhodes Fairbridge, a professor of geology at Columbia University, writes that this single volume, dealing with the *outer environment*, is designed to assist "all scientists, from those still in high school to the emeritus professor." The work is arranged alphabetically from *Aberration* to *Zone—climatic.* There are more than 150 contributors, and all articles are signed. Thoroughly illustrated; includes bibliographical references, a master index, and many cross-references.

3-d-2 Fairbridge, Rhodes Whitmore. *The Encyclopedia of Geochemistry and Environmental Sciences.* New York: Van Nostrand Reinhold, 1972. 1,321p.

This book is volume 4-A of the Encyclopedia of Earth Sciences series. Environmental science information has been added to a basic geochemistry reference book because of current interest and concern with the topic. The encyclopedia is alphabetical by key word or term with cross references from one article to another, and to articles in other volumes in the series where appropriate. References are appended to each article, and the preface contains a brief history of geochemistry. Indexed.

3-d-3 Firth, Frank E. *Encyclopedia of Marine Resources.* New York: Van Nostrand Reinhold, 1969. 740p.

In 1969, Frank Firth was associated with the University of Rhode Island's New England Marine Resources Information Program. According to Firth, this single-volume encyclopedia is a valuable tool for geologists, chemists, engineers, food technologists, metallurgists, fishery management specialists, and, more generally, "to all those who have any contact with, or interest in, the oceans and their almost infinite resources." The text is arranged alphabetically and each article is signed. There are 151 contributors. The work is liberally illustrated. An index and bibliographical references are provided. Most articles in the work include cross references.

3-d-4 McCrone, Walter C. *The Particle Atlas: A Photomicrographic Reference for the Microsiopical Identification of Particulate Substances.* Ann Arbor, MI: Ann Arbor Science, 1972. 406p.

Through photographic reproduction and measurement, this atlas serves as an index to environmental pollutants.

3-d-5 *McGraw-Hill Encyclopedia of Environmental Science.* Edited by Daniel N. Lapedes. New York: McGraw-Hill, 1974. 754p.

Most of the material in this volume has been published previously in the *McGraw-Hill Encyclopedia of Science and Technology*, 3d edition, 1971. This selection includes more than three hundred alphabetically arranged articles on topics from aerobiology to zooplankton. All articles are signed. A list of authors and their affiliations is included. Illustrated with black-and-white photographs and drawings. Comprehensive analytic index.

3-d-6 Nobile, Philip and Deedy, John, eds. *The Complete Ecology Fact Book*. Garden City, NY: Doubleday, 1972. 472p.

Divided into sections on population, endangered species, pollution, detergents, food, pesticides, nonrenewable mineral wastes, and solid wastes, this book attempts to bring together in a single volume statistics about ecology. Ample charts, maps, and diagrams illustrate topics while selected bibliographies are appended to each section. Appendixes contain information about notable court decisions and about foundations, government agencies, and other sources, as well as a glossary of terms. Indexed.

3-d-7 *North American Reference Encyclopedia of Ecology and Pollution*. Edited by William White, Jr. and Frank J. Little, Jr. Philadelphia: North American Publishing Co., 1972. 325p.

The third volume of the North American Reference Encyclopedia (previous volumes were devoted to women's liberation and drug abuse). This work consists of a series of essays of uneven quality on various aspects of ecology. Interesting short chapters may be found on "The Effects of Deicing Salt on Vegetation" and "Visual Pollution." A list of state environmental protection agencies is included. The Trispect Reference Index mixes organizations and monographic and periodical citations confusingly.

3-d-8 Todd, David Keith. *The Water Encyclopedia*. Port Washington, NY: Water Information Center, 1970. 559p.

A compendium of information about water in the United States and the world. Arranged in nine chapters, the encyclopedia covers such topics as water use, water quality, pollution control, and agencies and organizations involved with water-quality control. Full of climatic data, amply illustrated with charts and maps, and adequately indexed.

HANDBOOKS, GUIDES, AND MANUALS

3-e-1 *Advances in Ecology Research*, 1962–   . New York: Academic. Annual.

Designed for both general biologists and specialists in ecology. It includes four comprehensive and critical reviews of the significant literature of ecological research. Review articles contain extensive citations and bibliographies. There is a complete author and subject index to all citations.

3-e-2 *Advances in Environmental Science and Techology*, 1969–   . Edited by James N. Pitts and Robert L. Metcalf. New York: Wiley–Interscience. Irregular.

A series devoted to "the study of the quality of the environment and to the technology of its conservation." Contributions range from technical articles like "Catalytic Removal of Potential Air Pollutants from Auto Exhaust" to essays on general topics, such as "The Federal Role in Pollution Abatement and Control." Indexed by author and by subject.

3-e-3 *Annual Review of Ecology and Systematics*, 1970–   . Palo Alto, CA: Annual Reviews.

Averages sixteen reviews annually covering assessments of ecosystems, reproduction, aggression, resource management, systems analysis, human ecology, techniques, and instrumentation. Individual volumes are comprehensively indexed by author and subject.

3-e-4 CCM Information Corporation. *Environmental Pollution: A Guide to Current Research*. Produced from data gathered by Science Information Exchange, Smithsonian Institution, Washington, DC. New York: CCM Information, 1971. 851p.

A description of federally-sponsored research projects, compiled from data gathered in fiscal years 1969 and 1970. Divided into broad subject areas, the listings provide topics of investigation, names and addresses of investigators, descriptions of topics, and support institutions. Indexed by subject, researcher, and supporting agency.

3-e-5 *CRC Handbook of Environmental Control.* Edited by Richard G. Bond and Conrad P. Straub. Cleveland, OH: CRC Press, 1972–1973. 3 vols. 576p., 580p., 835p.

A three-volume set of which the first volume is devoted to environmental engineering, the second to the problems of collection, transportation, treatment, and disposal of solid wastes, and the third to water supply and treatment. Indexed.

3-e-6 *Ecology USA 1971: A History of the Year's Important Ecological Events.* New York: Special Reports, 1972. 619p.

The volume includes weekly reports of the important ecological events of the year. For each week, there is a résumé of events, a section on people in the news, a section on companies in the news, and a brief list of new books published on relevant topics. Name and subject indexes follow the weekly reports. The work also includes brief statements on ecology by major figures in government and business, the text of federal laws concerning the environment, a calendar of events, a title index to all books listed in the weekly reports, a table of acronyms, a list of state and federal government agencies involved in pollution control, and a lengthy directory of companies which manufacture or sell products, supply services, or perform research in the field of pollution control.

3-e-7 *Environmental Law Review*, 1970–   . Albany: Sage Hill. Annual.

A collection of the best law review articles which have a direct relationship to the practice of environmental law. The work contains subject and author indexes. Original-page citations to law review articles appear on each page of the text.

3-e-8 *Environment Regulation Handbook*, 1973–   . New York: Environment Information Center, Inc.

This subscription service provides chapters on the Environmental Protection Agency and the National Environmental Policy Act, as well as "regulatory" sections on air pollution, land use, mobile sources, noise, pesticides, radioactive materials, solid wastes, and water pollution. Indexed by subject and Standard Industrial Classification. A retrieval service provides limited-interest statutes upon request.

3-e-9 *Environment Reporter*, 1970– . Washington, DC: Bureau of National Affairs.

A subscription service which attempts comprehensive coverage of the environment, Federal and state developments are closely monitored, and the texts of all new environmental regulations and legislation can be found here. Coverage is topical in the Current Developments section, which also features brief coverage of coming events and governmental action. The monograph section offers topical essays such as "Federal Support for Environmental Research" and "Recycling Solid Wastes."

Contents as follows: Part 1, Federal Laws; Part 2, Federal Regulations; Part 3, State Air Laws; Part 4, State Water Laws; Part 5, State Solid Waste Land Use; Part 6, Monographs; Part 7, Current Developments; Part 8, Decisions/Cases.

3-e-10 *Environmental Law Reporter*, 1971– . Washington, DC: Environmental Law Institute.

A subscription service which reports on litigation involving the environment. Each issue is divided into Summary & Comments, Litigation, Administrative Proceedings, Statutory & Administrative Materials, Articles & Notes, and Indexes. The second section with texts of pending litigation is the longest. The Summary section contains a regular bibliography and news notes of environment-related litigation and developments. Articles & Notes is devoted to review articles of recent environmental litigation and legislation.

3-e-11 Landau, Norman J. and Rheingold, Paul D. *The Environmental Law Handbook*. New York: Ballantine, 1971. 496p.

Written in a popular style, this book is designed to present the legal remedies currently in existence to conserve and protect the environment. There are chapters on obtaining an attorney and on who may sue and who can be sued. Well indexed. The work also contains a technical glossary.

3-e-12 Noyes Data Corporation. *Pollution Analyzing and Monitoring Instruments—1972*. Park Ridge, NJ: Noyes, 1972. 354p.

A catalog, offering descriptions and specifications, of commercial equipment available for sampling, measuring, and analyzing environmental pollutants. Based upon manufacturers' information and specifications.

3-e-13 *Recognition of Air Pollution Injury to Vegetation: A Pictorial Atlas.* Edited by Jay S. Jacobson and A. Clyde Hill. Pittsburgh: Air Pollution Control Association, 1970. 112p.

Published as Information Report No. 1 of the TR-7 Agricultural Committee, Air Pollution Control Association, this book is the result of a three-year project which culminated in a symposium on the subject at the Association's annual meeting in 1969.

The atlas is organized into sections on each of the major phytotoxic air pollutants—ozone, sulphur dioxide, etc. Under each pollutant, the methods used to recognize and identify the resulting injury are described. Also included are: relative susceptibility of different species, pollutant concentration required to cause injury, and leaf analysis. The final chapter outlines an approach to recognizing injury, including methods for obtaining pertinent information, the value of field surveys, air monitoring, and chemical analysis. The book contains color photographs and an index of plant names.

3-e-14 *Your Government and the Environment*, 1971–   . Arlington, VA: Output Systems. Annual.

This publication includes chapters on air pollution, water pollution, radiation as a pollutant, pesticides, and solid wastes. Appendixes detail environmental spending, federal standards for environmental protection, state environmental agencies, and federal agencies concerned with the environment.

OTHER

3-f-1 Allerton Park Institute, 18th, 1972. *Information Resources in the Environmental Sciences: Papers Presented at the 18th Allerton Park Institute, November 12-15, 1972.* Edited by George S. Bonn. Urbana: University of Illinois Graduate School of Library Science, 1973. 238p.

Essays on resources in the federal government, the private sector, and national information centers, and on the informational activities of private and scientific organizations. An appendix contains a listing of agencies, services, and periodicals mentioned in the book.

3-f-2 Arny, Mary T. and Reaske, Christopher R. *Ecology: A Writer's Handbook.* New York: Random House, 1972. 112p.

A manual of composition intended for first-year college courses in English. The bibliography of ecology resources and the glossary of ecological terms make it a useful reference book for the individual who is concerned enough to write someone about the environment.

3-f-3 Montagu, Ashley. *The Endangered Environment.* New York: Mason & Lipscomb, 1974. 219p.

A book of quotations concerning eight subjects: air pollution, automobiles, water pollution, solid wastes, radiation, noise, DDT and other chemicals, and the SST. Quotations run from one line to several paragraphs. Each quotation is identified by source.

CHAPTER

HISTORIES

The history of ecology can be divided, bibliographically, into three areas of concern. The scientific introduction of the term *ecology* in the literature of biology and botany is well cataloged by John Richard Carpenter's *An Ecological Glossary* (see 3-b-1 for annotation), and there is really no need to look further, except in the very new disciplines like environmental chemistry, for the meaning of the concept and related concepts in scientific literature. For those who wish to explore these new disciplines and the use of ecology-oriented terminology in them, the authors have prepared a chart of monographs which claim to establish new disciplines (see Table 1).

The second area of concern is the growth of the concept of ecology out of the traditional discipline of conservation. There are a number of histories of conservation in America, some of which are listed below. There are also a few books devoted to the growth of ecology out of the conservation movement.

4-1 Clepper, Henry. *Leaders of American Conservation*. New York: Ronald, 1971. 353p.

A book that provides a "biographical recapitulation of the conservation movement in the United States, and of the development of professional resource management in the public interest." It consists of brief biographical descriptions of people who have contributed to the conservation of resources in the United States. Subjects were chosen by the member organizations of the Natural Resources Council of America. Each biographical statement is signed by an individual or organization, and many are reprinted from other sources. Most of the biographees are American citizens or residents; foreign nationals have been included only when their contributions to the field have been specifically related to conservation in the United States. This book is arranged alphabetically by surname of subject and is indexed by contributor.

Table 1

Reference	Author	Title	Field
5–5	Ager	Principles of Paleoecology	Paleoecology
5–96	Sondheimer and Simeone	Chemical Ecology	Chemical ecology
5–122	Dansereau	Biogeography	Biogeography
5–278	Lewis and Taylor	Introduction to Experimental Ecology	Quantitative ecology
5–369	Pielou	Introduction to Mathematical Ecology	Mathematical ecology
5–456	Patten	Systems Analysis and Simulation in Ecology	Systems ecology

4-2 Clepper, Henry and others. *Origins of American Conservation.* New York: Ronald, 1966. 193p.

A history of American conservation efforts by Clepper and eleven other authors. Wildlife, forests, fisheries, soil, water, range resources, parks and historic and scenic sites are all dealt with in separate chapters. Some historical portraits illustrate the book, and chapters offer suggestions for further reading. Indexed. The book contains a biographical section on contributors.

4-3 *Conservation in the United States: A Documentary History.* Edited by Leonard B. Dworsky. New York: Chelsea House in association with Van Nostrand Reinhold, 1971. 534p.

The first volume of this five-volume series on pollution reprints historical documents on air and water pollution. Brief histories introduce each section, and the entire volume carries an introduction by Stewart Udall. Since the other four volumes of the set are devoted to natural resources and recreation, libraries may want to acquire only the first volume in the series, unless they have strong conservation collections as well. Indexed.

4-4 Graham, Frank, Jr. *Man's Dominion: The Story of Conservation in America.* New York: Evans, 1971. 339p.

A history of the conservation movement in the United States from the middle 1880s until the passage of the Wilderness Act in 1964. Guy Bradley, Gifford Pinchot, John Muir, Steve Mather, William Temple Hornaday, Rosalie Edge, and Rachel Carson are all given prominent treatment. Indexed.

4-5 McHenry, Robert. *A Documentary History of Conservation in America.* Edited by Robert McHenry, with Charles Van Doren. Introduction by Lorus Milne and Margery Milne. New York: Praeger, 1972. 422p.

Arranged in four parts, this volume presents documented writings about conservation and related matters in the United States. In spite of some irrelevant material, and quite a lot of poetry, this is a useful collection. There is an index of authors, but little biographical information about the individuals anthologized. Not indexed.

4-6 Nash, Roderick. *The American Environment: Readings in the History of Conservation.* Reading, MA: Addison-Wesley, 1968. 236p.

A book of readings illustrating the American conservation movement in theory as well as practice. Sections on the conservation impulse, the Progressive conservation crusade, conservation between world wars, and conservation of the quality of the environment. The book contains a chronology of important conservation events and an extensive selective bibliography.

4-7 Smith, Frank Ellis. *The Politics of Conservation.* New York: Pantheon, 1966. 338p.

A history of conservation in which it is alleged that most conservation "victories" have their source in "pork barrel" politics. The author, a director of the Tennessee Valley Authority, was a congressman from Mississippi for twelve years and served on the House Public Works Committee. Indexed, with a lengthy bibliography.

4-8 Udall, Stewart L. *The Quiet Crisis.* Introduction by John F. Kennedy. New York: Holt, 1963. 209p.

This book, written by Stewart Udall, formerly Secretary of the Interior, is a history of the conservation movement in the United States. From Thomas Jefferson's land policy to the urbanization of America, the author documents the relationship of the people to the land and concludes with a statement covering conservation and the future. Illustrated with color and black-and-white photographs. Indexed.

Two collections of readings trace the growth of the present-day concern with ecology.

4-9 Worster, Donald E. *American Environmentalism: The Formative Period, 1860–1915.* New York: Wiley, 1973. 234p.

4-10 Pursell, Carroll W. *From Conservation to Ecology: The Development of Environmental Concern.* New York: T. Y. Crowell, 1973. 148p.

For a specific history of the Sierra Club, one should consult Holway R. Jones's *John Muir and the Sierra Club: The Battle for Yosemite* (5–251).

A final area of bibliographic concern is the documentation of Earth Day and its aftermath—when the environmental crisis became a major public issue. Many of the books listed in Chapter 5 record activities and developments since 1970. A particularly useful source is Anne Chisolm's *Philosophers of the Earth* (5–98), which provides a candid review of problems through interviews with leading ecologists. For information about Earth Day itself, the following book is helpful:

4-11 *Earthday—The Beginning: A Guide to Survival.* Compiled and edited by the national staff of Environmental Action. New York: Bantam, 1970. 233p.

The book contains a collection of speeches, by a long list of distinguished citizens on Earth Day, April 22, 1970. Contributions from politicians, political scientists, television commentators, and scholars are grouped together in ten subject areas. An appendix identifies the active regional groups of Environmental Action, organizers of the event.

CHAPTER

MONOGRAPHS

The publication of material in the area of ecology followed a rather traditional and logical pattern after Earth Day. Established writers in the broad areas of conservation, population biology, and general concern for the environment continued to publish substantial works. At the same time, the scholarly disciplines came to terms with the development of ecological concern on the part of the general public, and a number of college textbooks in botany and zoology began to reflect both the terminology and the concepts of ecology. Many new sub-disciplines took shape, as scholars in specific areas began to apply new concepts to traditional subjects. A particularly interesting development was the emergence of the text specifically devoted to an interdisciplinary course in ecology.

All the books which appear in this listing have been examined by the compilers. The titles are included not because they belong in every collection, but because each contains useful information. Indeed, the purpose of our indexing this material is to allow libraries to make a selection from the titles we have examined, and at the same time enable the reference librarian or the library patron to identify a particular book that might be of assistance in providing specific information.

Although most of the material included is not technical, some of it is. More than twice as many titles were examined as were included in this listing, and one guiding rule for selection was to include technical material only when necessary, when material of a more general nature did not provide a satisfactory substitute. Efforts were also made to balance the material, so a few classics were included, as well as some basic introductions to disciplines written from an ecological point of view. We hope it will prove a useful listing, and one which will provide assistance to those who seek information about the environment and the multitude of problems associated with it.

5-1 Abrahamson, Dean E. *Environmental Cost of Electric Power.* New York: Scientists' Institute for Public Information, 1970. 36p.

One of a series of nine pamphlet workbooks written by scientists for the layman and designed to explore topical environmental concerns. Other workbooks are devoted to air pollution, hunger, water pollution, environmental education, effects of weapons technology, pesticides, and nuclear explosives on the environment. Bibliography of suggested readings.

5-2 Adams, Alexander B. *Eleventh Hour: A Hard Look at Conservation and the Future.* New York: Putnam, 1970. 378p.

A popular monograph which suggests that conservation can "work." It contains chapters on the Everglades, dam builders, and the highway lobby. Appendixes contain notes on sources. Bibliography. Indexed.

5-3 Adelstein, Michael E. and Pival, Jean G., eds. *Ecocide and Population.* New York: St. Martin's, 1972. 120p.

A collection of readings on population growth and environmental decay. Each essay is accompanied by discussion and review questions. Bibliography.

5-4 Adler, Cy A. *Ecological Fantasies: Death from Falling Watermelons: A Defense of Innovation, Science, and Rational Approaches to Environmental Problems.* New York: Green Eagle, 1973. 350p.

The author attempts to debunk what he considers to be environmental fantasies. He deals specifically with "some of the more persistent environmental myths"—the air in the cities is poisonous, Lake Erie is dying, detergents are ecologically disastrous, and oil spills are permanently killing ocean life. Notes and references gathered at the end of the volume. Illustrated. Indexed.

5-5 Ager, Derek V. *Principles of Paleoecology*. New York: McGraw-Hill, 1963. 371p.

A text which explores ways of studying the ecology of prehistoric times from fossil groups and fossil assemblages. The author, an English geologist, draws heavily on the British Mesozoic and Cenozoic eras for illustration. The book contains a glossary of paleoecological terms, a lengthy bibliography, and an index with specific references to the illustrations.

5-6 *Air Pollution and Its Control*. Edited by R. W. Coughlin, A. F. Sarofim, and N. J. Weinstein. New York: American Institute of Chemical Engineers, 1972. 320p.

Technical papers presented at recent meetings of AICHE on ways to avoid and abate air pollution. The first section of the book provides an overview of combustion process research needs. Not indexed.

5-7 *Air Pollution Control*. Edited by Werner Strauss. New York: Wiley–Interscience, 1971. 2 vols. 451p., 300p.

These are two volumes in the Environmental Science and Technology Series of Texts and Monographs. The series editor teaches at the University of Melbourne, and Australian and British scholarship is heavily represented in this work. Part 1 is a collection of seven lengthy articles covering technical areas such as the control of sulfur emissions from combustion processes and collection of particles by fiber filters. Part 2 is a similar collection of five technical articles, including one essay on the literature of air pollution. Each article is exhaustive, covering theory, experimentation, and practice in the field. Charts and graphs. Both volumes indexed by author and subject.

5-8 *Air Pollution Control: Guidebook for Management*. Edited by A. T. Rossano, Jr. Stamford, CT: Environmental Science Service Division, E.R.A., 1969. 214p.

A practical text designed to provide the industrialist with a description and evaluation of air pollution problems based upon a "simplistic system model approach." Later chapters are devoted to source testing by stack sampling techniques and a case approach to successful air pollution abatement. A technical book whose final chapter and appendixes on the regulatory aspects of air pollution control are now somewhat dated. Not indexed.

5-9 Alaska Science Conference, 20th, University of Alaska, 1969. *Change in Alaska: People, Petroleum, and Politics.* Edited by George W. Rogers. College, AK: University of Alaska Press, 1970. 213p.

The proceedings of the 20th Alaska Science Conference with papers on the discovery of oil in Alaska and its attendant environmental problems. Sections are devoted to public policy for the management of petroleum resources and for the Alaskan environment. Indexed.

5-10 Aldous, Tony. *Battle for the Environment.* London: Fontana, 1972. 288p.

A book by the environment and architecture reporter of *The Times* on the challenges to Great Britain's new Department of the Environment. Chapters on automobile air pollution, London, conservation, industrial pollution, and planning highlight the book. Citizen involvement is urged. It contains a short bibliography. Indexed.

5-11 Allee, Warder C. and others. *Principles of Animal Ecology.* Philadelphia: Saunders, 1949. 837p.

A standard text in biology written from the standpoint of the ecologist. Sections on the history of ecology, analysis of the environment, populations, communities, and ecology and evolution. An extensive bibliography and author index. Also indexed by subject.

5-12 Allsopp, Bruce. *The Garden Earth: The Case for Ecological Morality.* New York: Morrow, 1972. 117p.

In this brief book a British professor of architecture considers the "political and moral implications of population and pollution." Calling for a new ecological morality, the author tries to "lay a few foundations upon which a new faith may be built." Indexed, with notes.

5-13 Altman, Philip L. and Dittmer, Dorothy S., comps. and eds. *Environmental Biology.* Bethesda, MD: Federation of American Societies for Experimental Biology, 1966. 694p.

This is an exhaustive volume of encyclopedic information about living organisms and their relationship to the environment. Designed for use by persons at all levels of biological study.

Sections on temperature, radiant energy, sound, vibration and impact, acceleration and gravity, atmosphere and pollutants, atmospheric pressures, gases, water solutes, and biological rhythms. An appendix contains tables of scientific and common names. Indexed.

5-14 American Assembly. *The Population Dilemma.* 2d ed. Edited by Philip M. Hauser. Englewood Cliffs, NJ: Prentice-Hall, 1969. 211p.

A book dealing with ecological aspects of population growth. Chapters on world population growth, population growth in developing countries, the adequacy of natural resources, population of the United States, population and economic development, problems of control, and policy issues. Indexed.

5-15 American Chemical Society. Committee on Chemistry and Public Affairs. Subcommittee on Environmental Improvements. *Cleaning Our Environment: The Chemical Basis for Action; a Report.* Washington, DC: American Chemical Society, 1969. 250p.

This publication presents "an objective account of the current status of the science and technology of environmental improvement" and "a number of measures that, if adopted, should help accelerate the sound development and use of that . . . technology." It is written for the layman and emphasizes the central role of chemistry in dealing with problems of the environment. It offers recommendations on the air environment, the water environment, solid waste, and pesticides. Indexed.

5-16 American Conference of Governmental Industrial Hygienists. Committee on Air Sampling Instruments. *Air Sampling Instruments for Evaluation of Atmospheric Contaminants.* 4th ed. Cincinnati: American Conference of Governmental Industrial Hygienists, 1972. var. paging.

Equipment and procedures for air sampling to isolate and identify pollutants in the air. Includes a glossary and index.

5-17 American Public Health Association. Program Area Committee on Air Pollution. *Guide to the Appraisal and Control of Air Pollution.* 2d ed. New York: American Public Health Association, 1969. 80p.

Written to assist the staff of a local agency assigned the responsibility for the monitoring and control of air pollution. Chapters are devoted to the role of the local agency, the estimation of the quantities of pollutants emitted, and the development of an agency program. Not indexed, but selected appendixes provide useful information; for example, estimates of particulates from industrial coal burning, according to firing method.

5-18 American Public Works Association. Committee on Solid Wastes. *Refuse Collection Practice.* 3d ed. Chicago: Public Administration Services, 1966. 525p.

A source book for local governments. Appendixes provide useful data on refuse collection practices in 956 American cities, typical refuse collection ordinances and regulations, a systems planning approach, and a selected bibliography. Indexed.

5-19 American Public Works Association. Institute for Solid Wastes. *Municipal Refuse Disposal.* 3d ed. Chicago: Public Administration Services, 1970. 538p.

A companion volume to *Refuse Collection Practice* (5-18), this book deals specifically with the various techniques for the disposal of refuse. Indexed, with an extensive bibliography.

5-20 American Water Works Association. *Water Q ality and Treatment: A Handbook of Public Water Supplies.* 3d ed. New York: McGraw-Hill, 1971. 654p.

Chapters are accompanied by extensive bibliographies and are devoted to such topics as filtration, radioactivity, and quality control in distribution systems. Indexed.

5-21 *America's Changing Environment.* Edited by Roger Revelle and Hans H. Landsberg. Boston: Houghton Mifflin, 1970. 314p.

Most of the essays in this book, which is volume 15 of The Daedalus Library, were originally published in the 1967 fall issue of *Daedalus*, the journal of the American Academy of Arts and Sciences. The book comprises articles by twenty economists, political scientists, city planners, and conservationists, who discuss the causes of the environmental problems facing the world, consider how serious they are, analyze what the general population thinks about them, and offer some ways of dealing with these problems through the American political system. Illustrated. Indexed.

5-22 Amidei, Rosemary. *Environment: The Human Impact.* Washington, DC: National Science Teachers Association, 1973. 252p.

Readings selected from *The Science Teacher* and arranged under the following topics: a point of view, aspects of the problem, environmental activities, and student activities. Illustrated. A brief annotated bibliography concludes the volume. Not indexed.

5-23 Amory, Cleveland. *Man Kind? Our Incredible War on Wildlife.* New York: Harper, 1974. 372p.

With grim humor, the author expresses anger and disgust at the destruction of wildlife and the cruelty of hunters. He blames state and federal game policies, hunting clubs, and even wildlife protection societies, many of which have lost sight of their original purpose. Amory points out groups still devoted to preservation and suggests ways in which citizens can help protect wildlife.

5-24 Anderson, Walt. *Politics and Environment: A Reader in Ecological Crisis.* Pacific Palisades, CA: Goodyear, 1970. 362p.

A collection of readings, designed for the college classroom, by well-known authors on the following topics: the politics of population, the varieties of pollution, four environmental crises, the urban environment, the rural environment, formulating environmental policy, and nature and human nature. Foreword by Senator Gaylord Nelson. Not indexed.

5-25 Andrews, William A., Moore, Donna K., and LeRoy, Alex. *A Guide to the Study of Environmental Pollution.* Englewood Cliffs, NJ: Prentice-Hall, 1972. 260p.

Part of the publisher's series entitled Contours: Studies of the Environment, this attractive textbook outlines basic problems of pollution and discusses the chemical and biological aspects of water pollution and air pollution. There are separate sections on research topics, field and laboratory studies, and case histories. Useful in upper-level high school classes or as a beginning college text. Well illustrated. Indexed.

5-26 *Are Our Descendants Doomed? Technological Change and Population Growth.* Edited by Harrison Brown and Edward Hutchings, Jr. New York: Viking, 1972. 377p.

Proceedings of a conference held at the California Institute of

Technology. This book is a collection of essays on the impact of technological change on the growth and concentration of human populations. Topics include developments in contraceptive devices, the role of the United Nations in population control, and the relation of the GNP to population growth and environmental damage. Indexed.

5-27 Armstrong, Terry R. *Why Do We Still Have an Ecological Crisis?* Englewood Cliffs, NJ: Prentice-Hall, 1972. 149p.

A book of readings with sections on values, economic and political strategies, technology, and education. It contains some original material. Not indexed.

5-28 Arthur, Don R. *Man and His Environment.* New York: American Elsevier, 1969. 218p.

Published in the United Kingdom under the title *Survival: Man and His Environment*, this book offers an interdisciplinary approach to human ecology. The author is a zoologist. There are sections on the origin of the human race, environment as a factor in survival, the effects of civilization on the environment, and compromises that may be necessary for survival. Illustrated with photographs, maps, and charts. Indexed.

5-29 Artin, Tom. *Earth Talk: Independent Voices on the World Environment.* New York: Grossman, 1973. 176p.

An account of the conference held concurrently with the United Nations Conference on the Human Environment and sponsored by the International Fellowship of Reconciliation (Dai Dong). An appendix contains this group's independent declaration on the environment. Not indexed.

5-30 Arvill, Robert. *Man and Environment: Crisis and the Strategy of Choice.* Rev. ed. Harmondsworth, Eng.: Penquin, 1969. 332p.

A book that depicts the environmental crisis in Great Britain. A program of resource planning is recommended and detailed in this book. Indexed with an extensive section of recommended reading. Illustrated with photographs and charts.

5-31 Ashby, Maurice. *Introduction to Plant Ecology.* 2d ed. New York: St. Martin's, 1969. 287p.

A botany text structured in terms of ecological concepts. Major sections of the text are devoted to tolerance, aggression, and plant communities. Since the book was first published in England, it has a British emphasis. Indexed, with an appendix on sampling soil atmosphere.

5-32 Asimov, Isaac. *ABC's of Ecology*. New York: Walker, 1972. 48p.

More advanced than its picture-book title would indicate, this book gives two entries related to ecology for each letter of the alphabet. Terms such as *biome*, *humus*, and *mutualism* are briefly explained. For each term there is an illustrative photograph or drawing. Grades 3–6.

5-33 *As We Live and Breathe: The Challenge of Our Environment.* Prepared by the Special Publications Division, National Geographic Society. Washington, DC: National Geographic Society, 1971. 239p.

A National Geographic book on the environment. Amply illustrated with color photographs, it has chapters devoted to agriculture, energy, industry, transportation, urbanization, waste, and population. Indexed.

5-34 Aylesworth, Thomas G. *This Vital Air, This Vital Water: Man's Environment Crisis*. Chicago: Rand McNally, 1968. 192p.

A young-adult title dealing with the environmental crisis. A short bibliography is included as well as a chapter on careers in the pollution-control field. Indexed. Grades 7+.

5-35 Bach, Wilfrid. *Atmospheric Pollution*. New York: McGraw-Hill, 1972. 144p.

This book is part of the McGraw-Hill Geography Series. The brief volume covers the origins and causes of air pollution, discusses its chemical, meteorological, health, economic, and technological aspects, and considers measures for air pollution control. A concluding section outlines ways in which the public can become involved in improving air quality. References cited at the end of each chapter. Not indexed.

5-36 Bain, Joe Staten. *Environmental Decay: Economic Causes and Remedies*. Boston: Little, Brown, 1973. 235p.

A book on "environmental economics" for the advanced undergraduate. Two chapters contain reprinted material, and each chapter contains a recommended list of additional readings. The first two chapters are devoted to a theoretical analysis of environmental degradation and its control, while chapters 4 through 8 use case histories to illustrate specific environmental problems. Water pollution, air pollution, the allocation of rural land and water resources, the death of the central cities, and adverse effects of freeways and air travel are explored. Not indexed.

5-37 *Balance and Biosphere: A Radio Symposium on the Environmental Crisis.* Toronto: Canadian Broadcasting Corporation, 1971. 113p.

The book contains the texts of seven radio talks first broadcast on the CBC program *Ideas* during the spring of 1970. The participants include Barry Commoner, K. E. F. Watt, Ivan Illich, W. A. Fuller, Paul Ehrlich, James Eayrs, and John Arapura. An annotated bibliography is also included. Not indexed.

5-38 Barnett, Harold J. and Morse, Chandler. *Scarcity and Growth: The Economies of Natural Resource Availability.* Baltimore: Johns Hopkins University Press (for Resources for the Future), 1965. 288p.

Historical and comparative economic analysis, structured around the growing scarcity of irreplaceable natural resources. Chapters on "contemporary views on the social aspects of natural resources" and the "conservation movement." Basic economic scarcity models are developed, and a series of tests proposed. Technical in nature. Indexed.

5-39 Baron, Robert Alex. *The Tyranny of Noise.* New York: St. Martin's, 1970. 294p.

A book about the environmental dimensions of noise, and the health and social costs of its presence in our environment. Bibliography. Indexed.

5-40 Barr, John. *The Assaults on Our Senses.* London: Methuen, 1970. 218p.

A thorough and factual survey of the environmental problems of Great Britain. Abuses are categorized according to the sense they offend—smell, sight, hearing, and so forth. Part five offers "Protests, Progress, Prospects." A list of organizations concerned with the environment and recommendations for further reading are also included. Effectively illustrated. Indexed.

5-41 Barros, James and Johnston, Douglas M. *The International Law of Pollution.* New York: Free Press, 1974. 476p.

Written by two Canadian professors, this is a college-level "collection of research and teaching materials." International agreements and judicial decisions are used as a basis for the book. Part 1 offers definitions, criteria, and priorities; part 2 covers major issues such as air pollution, international lakes and rivers, and environmental protection; part 3 "links specific pollution control data and more general questions of environmental protection," such as are raised by space exploration and nuclear weapons. Many chapters include lists of suggested readings. Indexed.

5-42 Bates, Marston. *The Forest and the Sea: A Look at the Economy of Nature and the Ecology of Man.* New York: Random House, 1960. 277p.

An examination of forms of life and their interrelationships in the sea. on a coral reef, and in the grasslands, deserts, tropical forests, and woods. Indexed, with a chapter on notes and sources.

5-43 Battan, Louis J. *The Unclean Sky: A Meteorologist Looks at Air Pollution.* Garden City, NY: Doubleday, 1966. 141p.

An introduction to the mechanisms and the hazards of air pollution. Written for the high-school student, the book deals in a concise way with "some chemistry of the atmosphere" and "the atmosphere as a dumping ground." Indexed.

5-44 Beatty, Rita Gray. *The DDT Myth: Triumph of the Amateurs.* New York: Day, 1973. 201p.

The author, a journalist, spent two years investigating conflicting claims about the dangers and benefits of DDT. Taking a strong position against "misguided environmentalists," the author's investigation supports continued use of DDT as an important insecticide which increases food production in a hungry world and has a reasonable safety hazard margin. The book concludes with the author's recommendations concerning the use of DDT. Selected list of references. Indexed.

5-45 Beckmann, Petr. *Eco-hysterics and the Technophobes.* Boulder, CO: Golem, 1973. 216p.

The author, a native of Czechoslovakia and now a professor of electrical engineering at the University of Colorado, calls his book a "doomsday debunker." His arguments are based on the proposition "that more, not less, science and technology is needed to eliminate pollution and clean up the environment. . . ." Attacking the "ecocult" for failing to do its homework, Beckmann does concede the existence of environmental problems and the limits of technology. Indexed.

5-46 Benarde, Melvin A. *Our Precarious Habitat.* Rev. ed. New York: Norton, 1973. 448p.

An examination of our environment as a system, with individual chapters on each of its major problems as determined by the author at the time of this book's revision. Well indexed. Each chapter carries recommendations for additional readings. A chapter on population has been added in this edition.

5-47 Bendick, Jeanne. *A Place to Live.* New York: Parents', 1970. 64p.

The simple but thoughtful text helps young readers become aware of their environment and their relationships to it. Illustrated with line drawings by the author, this book could be read to or by children with equal success. Indexed. Grades K-3.

5-48 Berger, Melvin. *Jobs That Save Our Environment.* New York: Lothrop, 1973. 96p.

One volume in the Exploring Careers series, this book describes jobs in the field of environmental control, including weather forecaster, geophysicist, oceanographer, plant ecologist, sewage plant worker, and city planner. Well illustrated with photographs which show women and members of minority groups in environmental careers. Includes indexes to careers and to career photographs. Grades 4-8.

5-49 Berger, Melvin. *The New Water Book.* New York: T. Y. Crowell, 1973. 111p.

This book combines the scientific study and analysis of water with the study of water pollution. The author begins with background on the properties of water, including simple experiments to illustrate those properties. He also discusses the pervasiveness of water, water sources, and distribution. Water pollution is explained in connection with specific rivers and lakes, and some simple experiments for the documentation of pollution are described. The author makes recommendations for student action and lists organizations students can contact for help. Illustrated with line drawings. Bibliography. Indexed. Grades 5-8.

5-50 Berger, Melvin. *Pollution Lab.* New York: Day, 1974. 125p.

This survey text explains how to detect pollution, purify air and water, and help protect the environment. Illustrated with black-and-white photographs. A list of suggested reading is included. Grades 6-8.

5-51 Berland, Theodore. *The Fight for Quiet.* Englewood Cliffs, NJ: Prentice-Hall, 1970. 370p.

Written for the layman, though packed with factual material, this book discusses what noise is and does, where noise comes from, and what the reader can do about it. Appended material includes a list of antinoise agencies in the United States; the author's noise exposure diary, in which he records noise levels in various situations; a list of principal sources; and a glossary. Indexed.

5-52 Berry, Wendell. *The Long-Legged House.* New York: Harcourt, 1965. 213p.

A poet's essays about strip mining, the despoiling of the environment, and the attitudes that cause people to perpetrate abuses upon nature. Not indexed.

5-53 Bews, John William. *Human Ecology.* New York: Russell & Russell, 1973. 312p.

First published in 1935 by Oxford University Press, this textbook on human ecology provides a historical introduction to the subject and deals in turn with anthropology, psychology, and sociology. A bibliography follows each chapter. Indexed.

5-54 Bigart, Robert. *Environmental Pollution in Montana.* New ed. Missoula, MT: Mountain Press, 1972. 261p.

Twenty-two essays reflecting opinions on all sides of the environmental pollution issue. Indexed, with appendixes on Montana legislation. Designed as a text for use in both high school and college.

5-55 Billings, W. D. *Plants, Man, & the Ecosystem.* 2d ed. Belmont, CA: Wadsworth, 1970. 160p.

An introduction to ecology by a botanist. The text, which is indexed, contains a glossary and suggestions for further reading.

5-56 *The Biological Impact of Pesticides in the Environment.* Edited by James W. Gillett. Corvallis: Oregon State University Press, 1970. 210p.

This book represents a symposium assessing the significance of pesticides in relation to the environment and health, held at Oregon State University in August 1969. Sections on the "transport and accumulation of pesticides in environments and ecosystems," the "effects of pesticides on fish," and the "effects of pesticides on health and neurophysiology of mammals." Appendixes contain biographical information about participants, a pesticide glossary and index, and an animal index. Most papers are technical in nature.

5-57 Bixby, William. *A World You Can Live In.* New York: McKay, 1971. 130p.

This book discusses the evolution of the ecological crisis, including the involvements of government, industry, and individual citizens in the problems and (potentially) in their solutions. The author includes suggestions for student action and a bibliography. Indexed. Grades 6-8.

5-58 Black, John N. *The Dominion of Man: The Search for Ecological Responsibility.* Chicago: Aldine, 1970. 169p.

A book of "applied ecology" that attempts to provide a historical perspective on the problems of our environment. The author, who is a professor of natural resources at the University of Edinburgh, has provided chapters on the western world view, concern for posterity, and the world of the future. A bibliography concludes the volume. Indexed.

5-59 Blau, Sheridan D. and Rodenbeck, John von B., comps. *The House We Live In: An Environment Reader.* New York: Macmillan, 1971. 515p.

Prepared for use in college classes in English composition where "provocative and carefully selected readings are required to provide rhetorical examples and to generate topics for student writing," this book has also been organized to be useful in interdisciplinary courses in environmental problems. The essays, many by familiar authors such as Wendell Berry, Rachel Carson, Paul Ehrlich, and René Dubos, is divided into a section on dangers and a section on causes and solutions. Illustrated.

5-60 Blaustein, Elliott H. *Anti-pollution Lab: Elementary Research, Experiments and Science Projects on Air, Water and Solid Pollution in Your Community.* New York: Sentinel, 1972. 128p.

Useful from the junior-high level on up, this book offers elementary procedures to locate pollution sources, measure the amount of pollution, examine the nature of the pollution, and gather information useful in fighting it. Covers air, water, and solid pollution. Illustrated. Appended are a list of questions to test the reader's knowledge of pollution, a brief bibliography, a list of activist organizations, and an index. Grades 7+.

5-61 Bohm, Peter and Kneese, Allen V., eds. *The Economics of the Environment.* New York: St. Martin's, 1971. 163p.

Reprinted from *The Swedish Journal of Economics*, this book presents technical information about measuring the cost of environmental degradation and the theory and policy of environmental management. Not indexed.

5-62 Boughey, Arthur S. *Contemporary Readings in Ecology.* Belmont, CA: Dickenson, 1969. 390p.

Twenty-two technical papers, reprinted from other sources and grouped under the following topics: taxonomy, evolutionary ecology, population ecology, and community ecology. Designed as supplementary reading for a college course in ecology, this book is useful because of its conceptual organization. Not indexed.

5-63 Boughey, Arthur S. *Ecology of Populations*. 2d ed. New York: Macmillan, 1973. 182p.

Chapters are devoted to the environmental requirements of organisms, the behavior of populations, population interaction, population evolution, the formation of communities, and human ecology. Each chapter offers further recommended readings. Indexed.

5-64 Boughey, Arthur S. *Fundamental Ecology*. Scranton, PA: Intext, 1971. 222p.

This volume is the introductory text to the publisher's Series in Ecology, which includes monographs on specialized subjects in the field of ecology. This book is designed for use by undergraduates in introductory courses or by laymen. It covers the definition and history of ecology; ecosystems; environments; population dynamics, evolution, interaction, and behavior; and the nature of communities. Each chapter includes a bibliography and list of review questions. Glossary. Illustrated. Indexed.

5-65 Boughey, Arthur S. *Man and the Environment: An Introduction to Human Ecology and Evolution*. 2d ed. New York: Macmillan, 1975. 472p.

This book is designed as an introductory college text. The author begins with a discussion of basic ecological constructs, then provides an "ecological account of the origins and final emergence of our urban technocracy," and finally deals with individual problems such as population control, air and water pollution, pesticides, and conservation. Each chapter includes a list of references and a list of further readings. Appended material includes a glossary, a list of governmental regulatory agencies, and information on geological dating. Separate author and subject indexes.

5-66 Boughey, Arthur S. *Readings in Man, the Environment, and Human Ecology*. New York: Macmillan, 1973. 595p.

This collection of readings was compiled by Arthur Boughey for use with his text *Man and the Environment: An Introduction to Human Ecology and Evolution*. The two books are parallel in organization. The selections reprinted here are intended to represent current thinking on such major topics as the ecosphere, human evolution, the origins of society, demography and population density, pollution and environment, and conservation and the future. Each section also contains a list of additional readings. Glossary included. Illustrated, primarily with charts and graphs. Separate author and subject indexes.

5-67 Bova, Benjamin. *Man Changes the Weather.* Reading, MA: Addison-Wesley, 1973. 159p.

The author describes the ways in which man has changed the world's atmosphere. He includes planned changes such as rainmaking and accidental changes due to pollution. Appended is a list of government weather modification programs. Indexed. Grades 5-8.

5-68 Boyle, Robert H.; Graves, John; and Watkins, T. H. *The Water Hustlers.* New York: Sierra Club, 1971. 253p.

Three regional water plans, in Texas, California, and New York, are attacked by the authors. The section by John Graves on the Texas plan is the largest. Not indexed.

5-69 Bragdon, Clifford R. *Noise Pollution–The Unquiet Crisis.* Philadelphia: University of Pennsylvania Press, 1970. 280p.

An exhaustive examination of noise as an element of environmental concern. Chapters are devoted to the nuisances and hazards of noise and to the design and analysis of the Metropolitan Philadelphia Noise Survey. A lengthy bibliography is contained in an appendix. Indexed.

5-70 Brainerd, John W. *Working With Nature: A Practical Guide.* New York: Oxford, 1973. 517p.

A text, arranged around working experiments in nature, that offers high-school and beginning college students some insight into the functioning of the natural environment. Divided into sections on typical environments (drylands, for example) and environmental activities (such as soil management), the text contains numerous cross references between different sections. Illustrated with an extensive and useful bibliography. Indexed.

5-71 Brehman, Thomas R. *Environmental Demonstrations, Experiments and Projects for the Secondary School.* West Nyack, NY: Parker Publishing, 1973. 213p.

Environmental projects and models presented around the concepts of land use planning and water and natural resource utilization. Each section recommends classroom materials and a list of further readings. Indexed.

5-72 Bresler, Jack Barry. *Environments of Man*. Reading, MA: Addison-Wesley, 1968. 289p.

Collected readings on the environment, with sections on adaption to soils and physical factors. Additional sections on food, water and energetics, reproduction, cycles, ecology and genetics, population and society, pollution and hereditary consequences. A final section is devoted to speculations about life on Mars, based on what is known about the Martian environment. Appendixes contain a glossary of terms and a supplementary annotated bibliography. Not indexed.

5-73 Brittin, Wesley Emil and others, eds. *Air and Water Pollution; Proceedings of the Summer Workshop, August 3 to August 15, 1970, Universtiy of Colorado*. Boulder, CO: Associated University Press, 1972. 613p.

Fourteen technical papers on water pollution and nineteen technical papers on air pollution present the results of this symposium. Most papers are structured to present "the most important unanswered basic research problems." The Water Pollution Workshop dealt with effects of water pollutants on biota, physical and chemical phenomena in water, and waste-water treatment. The Air Pollution Workshop dealt with problems in atmospheric chemistry, local pollution dispersion studies, global effects of air pollution, and effects of air pollution on biota. Not indexed.

5-74 Brooks, Paul. *The Pursuit of Wilderness*. Boston: Houghton Mifflin, 1971. 220p.

This book deals with key battles for the preservation of wilderness, all but one of which were fought in the five years before the book's publication. Essays about the Northwest, Alaska, and Florida. The fate of East African game herds and the preservation of the rhinoceros are also treated. Illustrated with photographs. Not indexed.

5-75 Brower, David. *Wildlands in Our Civilization*. New York: Sierra Club, 1964. 175p.

Selected papers with accompanying photographs from the first five biennial Wilderness Conferences sponsored by the Sierra Club. Summaries of the first five conferences are reprinted in part 2 of this book. Not indexed.

5-76 Brubaker, Sterling. *To Live on Earth: Man and His Environment in Perspective.* Baltimore: Johns Hopkins University Press (for Resources for the Future), 1972. 202p.

A study which deals with man's accelerating use of the environment, environmental threats, coping with environmental problems, and some "larger considerations." Global climate, radioactivity, pesticides, fertilizers, and erosion are treated. References follow each chapter. The book contains a short bibliography. Indexed.

5-77 Bryerton, Gene. *Nuclear Dilemma.* New York: Ballantine, 1970. 138p.

An examination of nuclear power and its attendant problems as opposed to energy derived from fossil fuels. Chapters on the dangers of reactor malfunction, biological hazards, thermal pollution, and ecological effects. Two chapters are devoted to power alternatives and the "critics' case." Illustrated, with a lengthy bibliography. David R. Inglis prepared the foreword.

5-78 Bugler, Jeremy. *Polluting Britain: A Report.* Harmondsworth, Eng.: Penguin, 1972. 182p.

A summary of pollution in Great Britain, by the environmental correspondent for the *Observer.* Chapters on sound, rivers, the sea, parks, and the landscape. Tougher enforcement of antipollution regulations is urged, following the trends in the United States. Indexed.

5-79 Burch, William R., Jr. *Daydreams and Nightmares; A Sociological Essay on the American Environment.* New York: Harper, 1971. 175p.

An extended essay by a sociologist which explores "some ways in which myth, social structure, and ecosystem interpenetrate." There are chapters on the "rhetorical uses of nature" and "democracy and conservation." Of more interest to sociologists than ecologists. Indexed.

5-80 Burch, William R., Jr.; Cheek, Neil H., Jr.; and Taylor, Lee. *Social Behavior, Natural Resources, and the Environment.* New York: Harper, 1972. 374p.

A group of original essays by sociologists on the "unfolding relations between social behavior, natural resources, and envi-

ronment." Essays include a comparison of the civil rights movement and the environmental movement and an analysis of the
"logic" of current studies of human ecology. Indexed.

5-81 Burns, William. *Noise and Man*. Philadelphia: Lippincott,
1969. 336p.

An examination of noise and its physiological effects upon
human beings by an English physiologist. A chapter is devoted to
the effects of noise on health and to the permanent effects of noise
on hearing. An appendix contains a glossary of acoustical terms.
Indexed by subject and author.

5-82 Calder, Nigel. *Eden Was No Garden: An Inquiry into the
Environment of Man*. New York: Holt, 1967. 240p.

A speculative treatise on human ecology, offering suggestions
and scenarios to deal with future energy and territorial requirements. Not indexed.

5-83 Caldwell, Lynton Keith. *Environment: A Challenge for Modern Society*. Garden City, NY: Natural History, 1970. 292p.

According to the author, "the message of the book is that if
modern man and his civilization are to survive, administration of
man's environmental relationships must become a major task of
government." The book is divided into three major sections:
policy, tasks, and management. It is written for the layman.
Indexed, with extensive chapter notes.

5-84 Caldwell, Lynton Keith. *In Defense of Earth: International
Protection of the Biosphere*. Bloomington: Indiana University
Press, 1972. 295p.

A history of international concern with the environment. Appendixes carry a list of international organizations and programs
concerned with environmental issues and a list of international
conferences and agreements for the protection of the biosphere.
Indexed.

5-85 Campbell, Rex R. and Wade, Jerry L. *Society and Environment: The Coming Collision*. Boston: Allyn & Bacon, 1971. 376p.

A book of readings compiled by two sociologists on the collision
referred to in the title, water, air, the problem profusion, population, and social transformation. Each of the six sections carries a
bibliography for further reading.

5-86 Caras, Roger. *Last Chance on Earth: A Requiem for Wildlife.* New ed. New York: Schocken, 1972. 207p.

A status report on forty endangered species, with illustrations by Charles Fracé. The preface to this edition contains a status report on each of the animals since the book's original publication in 1966. The book includes a "selected bibliography" and a list of conservation organizations.

5-87 Carr, Donald Eaton. *The Breath of Life.* New York: Norton, 1965. 175p.

A book by a research chemist on the automobile's effect on the atmosphere. The focus is on California, particularly Los Angeles. The electric car is suggested as the most feasible solution to the problem of cleaning up the atmosphere. Not indexed.

5-88 Carr, Donald Eaton. *The Deadly Feast of Life.* Garden City, NY: Doubleday, 1971. 369p.

The author examines food and eating, how species become extinct, and the ways animals use poisons. By carefully observing the whole animal kingdom and past evolutionary patterns, he hopes to gain insight into the future of the human species. The last several chapters deal specifically with man-made poisons, such as pesticides and food additives. Appended are charts of geologic time, a diagram of evolutionary development of the species, and an example of animal classification. Bibliography included. Indexed.

5-89 Carr, Donald Eaton. *Death of Sweet Waters.* New York: Norton, 1971. 257p.

An examination of the various problems and influences which affect adequate supplies of potable water. Soil management, river basin behavior, and patterns of rainfall are all touched upon. Indexed, with an extensive bibliography.

5-90 Carson, Rachel. *Silent Spring.* Boston: Houghton Mifflin, 1962. 368p.

One of the first books to direct popular attention to the problems of the environment. It concerns a broad range of topics and should be primary reading for those concerned with documenting the environmental crisis. An appendix contains a list of principal sources. Indexed.

5-91 Carvell, Fred J. and Tadlock, Max, comps. *It's Not Too Late.* Beverly Hills, CA: Glencoe, 1971. 312p.

A book of readings about ecology, compiled by urban planners and arranged around five topics: the nature and scope of ecological concern, change in the environment, signs of hope, the need to establish priorities, and the philosophical prelude to environmental action. Not indexed.

5-92 Caudill, Harry. *My Land Is Dying.* New York: Dutton, 1971. 144p.

This book deals with strip mining and offers poignant photographic documentation of its effects on the environment. Illustrations are primarily from Appalachia, and specifically the author's Kentucky, but strip mining as a national plague is detailed in this book.

5-93 Chanlett, Emil T. *Environmental Protection.* New York: McGraw-Hill, 1973. 569p.

A college textbook on environmental management and protection. The material is "man-centered," according to the author. It is also fairly technical and requires serious concentration. Dealing in separate chapters with water resources, air pollution, solid waste management, food protection, and so on, the author examines the consequences of mismanagement at three levels: effects on health; effects on comfort, convenience, and efficiency; and effects on the balance of ecosystems and renewable resources. References are listed at end of each chapter. Appended material includes a list of selected infectious diseases and their environmental agents, the electromagnetic spectrum, and the sonic range. Indexed.

5-94 Chase, Stuart. *Rich Land, Poor Land: A Study of Waste in the Natural Resources of America.* New York: McGraw-Hill (Whittlesey House), 1936. 361p.

An early book that catalogs environmental disasters across the United States. The entire range of natural resources is dealt with. An important historical catalog for the environmental collection. Illustrated by photographs. Indexed.

5-95 Chedd, Graham. *Sound, from Communication to Noise Pollution*. Garden City, NY: Doubleday, 1970. 187p.

An interesting, well-illustrated book, this volume carefully presents the basic principles of sound, speech, hearing, acoustics, ultrasound, and noise. The final chapter on noise control is more meaningful because of the foundation established in earlier chapters. Many color photographs and illustrations. Indexed.

5-96 *Chemical Ecology*. Edited by Ernest Sondheimer and John Simeone. New York: Academic, 1970. 336p.

A text that deals with the chemistry of ecological interaction in plants, insects, fish, and arthropods. This book grew out of a series of lectures given in 1968 at the State University of New York College of Forestry at Syracuse University. Indexed by subject and author.

5-97 *Chemical Mutagens: Principles and Methods for Their Detection*. Edited by Alexander Hollaender. New York: Plenum, 1971-1973. 3 vols. 304p., 300p., 310p.

This work grew out of a symposium sponsored by the Environmental Mutagen Society. The material is technical in nature. Indexed by subject and author.

5-98 Chisholm, Anne. *Philosophers of the Earth*. New York: Dutton, 1972. 201p.

A book by an English journalist, based upon interviews with a number of ecologists. The author's purpose is to allow "a selection of the people whose activities and ideas have formed and are forming ecology to explain, in their own way, the range and substance of an extraordinarily varied, complex, and crucial subject." Interviews with Barry Commoner and Paul Ehrlich are featured. Not indexed.

5-99 Chute, Robert M. *Environmental Insight: Readings and Comment on Human and Nonhuman Nature*. New York: Harper, 1971. 241p.

Meant to provide undergraduates who are not majoring in science with "a greater insight into the relationship of man to nonhuman nature," this text treats a broad spectrum of environmental concerns from a biological point of view. Topics include ecosystems, the effects of various toxic substances in nature, the hazards of radioactive waste, population dynamics, and so on. Each essay is reprinted *in toto* and is accompanied by editorial comments from the compiler. Most of the selections carry lists of references or a bibliography. Indexed.

5-100 Clarke, George L. *Elements of Ecology.* Rev. ed. New York: Wiley, 1965. 560p.

A reprint of the original 1954 textbook, with a lengthy new bibliography to bring it up to date. Indexed.

5-101 *Coastal Zone Resource Management.* Edited by James C. Hite and James M. Stepp. New York: Praeger, 1971. 169p.

One of the Praeger special studies in American economic and social development, this monograph consists of several scholarly essays which "discuss a wide range of issues related to developing a scientific management system for the Coastal Zone." The editors feel that political institutions are the ultimate channel for decisions concerning coastal management. Each essay concludes with notes. Appended material includes changes in water permit regulations, descriptions of illustrative state programs of estuarine conservation, and notes on contributors. Not indexed.

5-102 Coastal Zone Workshop, Woods Hole, MA, 1972. *The Water's Edge.* Cambridge: MIT Press, 1972. 393p.

The proceedings of a workshop held from May 22 to June 3, 1972, and sponsored jointly by the Institute of Ecology and the Woods Hole Oceanographic Institution. The first chapter contains the workshop's conclusions and recommendations. The development of a comprehensive National Coastal Zone Policy is recommended. Part 2 of the book has to do with restraints on human activities in order to protect the environment, and part 3 is devoted to the management of coastal resources. Comprehensive in scope, the book has a short bibliography and is indexed by author and subject. The participants in the workshop are listed in the preface.

5-103 Coblentz, Stanton A. *The Challenge to Man's Survival.* South Brunswick, NJ: A. S. Barnes, 1972. 292p.

An examination of human chances for survival in the age of technology, with references drawn from both the natural and social sciences. One chapter is devoted to the United States;

other chapters deal with worldwide problems caused by pesticides and pollution. Indexed.

5-104 Colinvaux, Paul A. *Introduction to Ecology.* New York: Wiley, 1973. 621p.

A general introduction for the college student. The book is organized into four parts: part 1 examines the emergence of ecology from biogeography; part 2 examines the ecosystem concept; part 3 deals with "the balance of nature," and part 4 is a review and synthesis. Illustrated. Extensive bibliography, and a combined glossary and index.

5-105 Commoner, Barry. *The Closing Circle: Nature, Man, and Technology.* New York: Knopf, 1971. 326p.

Case studies of nuclear, air, and water pollution document the author's thesis—that technology has endangered the vital balance of nature. Indexed, with an extensive section of notes by the author.

5-106 Commoner, Barry. *Science and Survival.* New York: Viking, 1963. 150p.

A significant early work by one of the most important environmental authors. Stressing the social aspects of science, Commoner deals with most contemporary environmental problems and concludes, "Science can reveal the depth of the crisis, but only social action can resolve it." Indexed.

5-107 Conference on Law, Science and Politics; water pollution and its effects considered as a world problem, Aberystwyth, 1970. *Water Pollution as a World Problem: The Legal, Scientific, and Political Aspects.* Report of a conference held at the University of Wales, Aberystwyth, July 11-12, 1970. London: Europa (for the David Davies Memorial Institute of International Studies), 1971. 240p.

Examines water pollution from nuclear and thermal wastes, oil, chemicals and pesticides, industrial by-products, and sewage. Although the participants were all from the United Kingdom, the essays range far and wide. Appendixes carry the text of Richard Nixon's Message on Ocean Dumping, the text of Canadian legislation to control water pollution in the Arctic, and the Convention text on the protection of wetlands of international importance. Not indexed.

5-108 Conference on Legal and Institutional Responses to Problems of the Global Environment, Arden House, 1971. *Law, Institutions, and the Global Environment.* Edited by John Lawrence Hargrove. Dobbs Ferry, NY: Oceana, 1972. 394p.

Proceedings of the conference, as well as additional essays which introduce portions of the monograph and summarize discussions. The Conference was sponsored by the Carnegie Endowment for International Peace and the American Society of International Law. Participants included "legal advisors and other officials of international organizations . . . , certain government officials, and members of the academic community." The book is divided into four sections: a framework of policy issues, the international interest in the environment, toward an international law of environmental protection, and developing institutional processes and structures. Indexed.

5-109 Conference on the Ecological Aspects of International Development, Airlie House, 1968. *The Careless Technology: Ecology and International Development; the Record.* Edited by M. Taghi Farvar and John P. Milton. Garden City, NY: Natural History, 1972. 1,030p.

The proceedings of a conference on ecology and international development. Technical papers are grouped under five topics: health and nutritional consequences of selected development programs; irrigation and water development; ecological consequences of intensification of plant productivity; intensification of animal productivity; and special problems of environmental degradation. A final section is devoted to the implications of the conference for international development programs. Includes an index of scientific names and biographical information about the participants.

5-110 Conference on the Ecology of Human Living Environments, 1st., Green Bay, Wisconsin, 1971. *Behavior, Design, and Policy: Aspects of Human Habitats.* Edited by William M. Smith. Green Bay, WI: Office of Community Outreach, University of Wisconsin—Green Bay, 1972. 228p.

This conference brought together experts in the fields of human behavior, architecture, regional and urban planning, and medicine to discuss man and his relation to habitats. There are thirteen papers; general discussion is recorded at the end of each paper and in a final section of the book. Biographical notes on contributors conclude the volume. Not indexed.

5-111 *Congress and the Environment.* Edited by Richard A. Cooley and Geoffrey Wandesforde-Smith. Seattle: University of Washington Press, 1970. 277p.

Sixteen essays which examine the role Congress has played in dealing with the problems of our environment. Treated are the Indiana Dunes National Lakeshore, The Highway Beautification Act, The Wilderness Act, wild and scenic rivers, aircraft noise abatement, and a host of other topics. Introductory and concluding essays provide focus for this collection. Indexed, with a useful bibliography.

5-112 Congressional Quarterly, Inc. *Man's Control of the Environment.* Edited by Jack Nease. Washington, DC: Congressional Quarterly, Inc., 1970. 91p.

An analysis of environmental legislation and progress, as of August 1970, by the staff of *Congressional Quarterly.*

5-113 Connell, Joseph H.; Mertz, David B.; and Murdoch, William W., eds. *Readings in Ecology and Ecological Genetics.* New York: Harper, 1970. 397p.

Designed to introduce the beginning college student to the literature of ecology. Each section offers an overview of material presented. There are sections entitled "Life Histories and Ecological Genetics," "Distribution and Abundance," and "Ecological Communities." The journal articles reprinted are from magazines like *Ecology, Evolution,* and *Researches on Population Ecology.* Not indexed.

5-114 Conservation Foundation. *National Parks for the Future.* Washington, DC: Conservation Foundation, 1972. 254p.

Recommendations which resulted from a symposium held on national parks in April 1972. The book consists of the Conservation Foundation's overview and recommendations, results of the symposium's task force reports, reports presented at the Yosemite Symposium, and selected project papers. The purpose of this book is to provide environmentally sound recommendations for the next hundred years of the National Park System. Not indexed.

5-115 Cottam, Walter P. *Our Renewable Wild Lands–A Challenge.* Salt Lake City: University of Utah Press, 1961. 182p.

A series of essays on conservation problems in Utah's wild lands. Titles include "Is Utah Sahara Bound?" and "A Comparative Study of Vegetation." Illustrated with photographs. Bibliographies. Indexed.

5-116 Crenson, Matthew A. *The Un-Politics of Air Pollution: A Study of Non-Decision-Making in the Cities.* Baltimore: Johns Hopkins University Press, 1971. 227p.

Case studies of Gary, Indiana and East Chicago, Illinois. Chapters on the influence of industry on regulation and the politics of pollution control and abatement. An appendix contains a National Opinion Research Center Interview Schedule for a panel of community leaders. Indexed.

5-117 *The Crisis of Survival.* By the editors of *The Progressive* and the College Division of Scott, Foresman and Company. With introductions by Eugene P. Odum and Benjamin DeMott. New York: Morrow, 1970. 261p.

Reprinted material by distinguished writers which appeared, for the most part, in a *Progressive* special issue entitled "The Crisis of Survival" (April 1970). Not indexed, but the book contains a brief bibliography.

5-118 Crocker, Thomas D. and Rogers, A. J. III. *Environmental Economics.* Hinsdale, IL: Dryden, 1971. 150p.

A short monograph printed entirely on reclaimed paper, that attempts to demonstrate that environmental quality is a "scarce good." Although intended as a college text, it is written in a manner which seems condescending to college students. It might, however, be useful in the collection for high-school students, for it does provide an introduction to economics, as well as an introduction to the problems of our environment. A list of recommended readings is provided. Not indexed.

5-119 Crowe, Philip K. *The Empty Ark.* New York: Scribner, 1967. 301p.

This is an account of three major journeys undertaken by the author on behalf of the World Wildlife Fund. Traveling to East Africa and the Middle East in 1963, South America in 1964, and East Asia and the Pacific in 1965, the author surveyed laws protecting rare species, the conditions and probable future of those species, and their habitats. Illustrated with black-and-white photographs. Indexed.

5-120 Curry-Lindahl, Kai. *Conservation for Survival: an Ecological Strategy.* New York: Morrow, 1972. 335p.

This book deals with the renewable natural resources (air, water, soil, vegetation, and animals) and the necessity of conservation for survival. Both the sea and fresh water are dealt with, and there is a separate chapter on human beings. Problems existing on all continents are treated; the ecological problems of the "Green Revolution" are considered in a chapter on the future. The book is a comprehensive argument for conservation and is filled with recommendations and proposals on a myriad of subjects. Glossary and bibliography. Indexed.

5-121 Curry-Lindahl, Kai. *Let Them Live: A Worldwide Survey of Animals Threatened With Extinction.* New York: Morrow, 1972. 394p.

Organized by geographical location. Bibliography. Indexed.

5-122 Dansereau, Pierre. *Biogeography: An Ecological Perspective.* New York: Ronald, 1957. 394p.

A textbook designed to provide the advanced college student with a "new synthesis of the environmental relationships of human organisms." An examination of the "ecological processes" that are at work in various natural environments. Glossary and an extensive bibliography. Indexed.

5-123 Darling, Frank Fraser. *Wilderness and Plenty.* Boston: Houghton Mifflin, 1970. 84p.

The Reith lectures for 1969, with essays on man and nature, the impact of man on the environment, and forward vision in conservation. The Reith lectures are a series of radio talks commissioned by the British Broadcasting Corporation. Bibliography. Not indexed.

5-124 Darling, Frank Fraser and Milton, John P. *Future Environments of North America: Transformations of a Continent.* Garden City, NY: Natural History, 1966. 767p.

The proceedings of a conference sponsored by The Conservation Foundation in April 1965. Presentations and general discussion are grouped into the following categories: the organic world and its environment; regions—their developmental history and future; economic patterns and processes; social and cultural purposes; regional planning and development; and organizations and implementation. Closing statement by Lewis Mumford. Most articles carry bibliographies. Indexed.

5-125 Darling, Lois and Darling, Louis. *A Place in the Sun: Ecology and the Living World*. New York: Morrow, 1968. 128p.

The authors detail basic ecological principles and stress the need for people to act more carefully, lest they do irreparable harm to their environment and thus to themselves. A discussion of environmental peril goes beyond such topics as dangerous pesticides to the possibilities of glacial thaw and oxygen shortage. Includes a substantial selected bibliography. Indexed. Grades 6-9.

5-126 Dasmann, Raymond F. *The Destruction of California*. New York: Macmillan, 1965. 247p.

A book on the environmental disasters that have befallen the state of California, partly because of neglect. Sections on water, air, timbering, farmland, and planning. A most thorough treatment. Illustrated with maps and photographs. A bibliography of references is included. Indexed.

5-127 Dasmann, Raymond F. *A Different Kind of Country*. New York: Macmillan, 1968. 276p.

The book makes a plea for diversity. It is about the American wilderness, primarily, though it also deals with urban life. Indexed, and amply illustrated with photographs.

5-128 Dasmann, Raymond F. *Environmental Conservation*. 3d ed. New York: Wiley, 1972. 473p.

This book, written for the beginning university student and the general public, is intended to serve as the text for a one-semester course concerned with man and the environment. In this edition, increased attention is given to problems of the global environment, though most of the material pertains to the North American continent. Illustrated with maps, charts, and photographs. Indexed.

5-129 Dasmann, Raymond F. *No Further Retreat: The Fight to Save Florida*. New York: Macmillan, 1971. 244p.

Threatened areas are examined in detail, and efforts towards conservation of unique habitats are described. Bibliography. Indexed.

5-130 Dasmann, Raymond F. *Planet in Peril: Man and the Biosphere Today.* New York: World, 1972. 242p.

The author, senior ecologist for the International Union for the Conservation of Nature and Natural Resources, surveys the environmental crisis, the workings of the biosphere, the impact of civilization on the environment, and the conservation movement. Finally, he discusses possible alternative goals the human race may wish to attain. This is a UNESCO book. Illustrated with black-and-white photographs. Bibliography. Indexed.

5-131 Dasmann, Raymond F.; Milton, John P.; and Freeman, Peter H. *Ecological Principles for Economic Development.* New York: Wiley, 1973. 252p.

This book was written "for the use of those concerned with development, whether at a purely national level or in connection with any of the aid programmes of . . . international agencies." It grew out of a conference on the ecological aspects of international development (see 5-109, *The Careless Technology*), and deals with humid tropical lands, tourism, agricultural development projects, and river basin development projects. Indexed.

5-132 Davies, J. Clarence. *The Politics of Pollution.* New York: Pegasus, 1970. 231p.

The author, who spent two years with the U.S. Bureau of the Budget as examiner for environmental and consumer protection, describes the pollution problem and the growth of interest in environmental quality. In the second part of his book, the forces which have shaped pollution policy are examined. A final section examines the major policies for controlling pollution, and the last chapter makes several recommendations for their improvement. Indexed.

5-133 Degler, Stanley E. *Oil Pollution: Problems and Policies.* Washington, DC: Bureau of National Affairs, 1969. 142p.

This is a collection of several articles and documents on oil pollution, including a paper on oil pollution and the law, spillage prevention, control, and restoration; a report on oil pollution to the President by the Secretary of the Interior and the Secretary of

Transportation; excerpts from a multi-agency contingency plan; and the texts of the Oil Pollution Act of 1924 and the Oil Pollution Act of 1961. The first chapter also includes a list of emulsifiers for oil dispersal and the companies that produce them. Not indexed.

5-134 Denny, Ronald C. *This Dirty World.* London: Nelson, 1971. 211p.

Although the author is British, this book on pollution is not restricted solely to the United Kingdom. The United States and the high seas are also treated. An appendix contains the names and addresses of organizations in the United Kingdom concerned with environmental pollution. Extensive notes and bibliography are contained in a separate section of the book. Indexed.

5-135 Detweiler, Robert; Sutherland, Jon N.; and Werthman, Michael S., eds. *Environmental Decay in Its Historical Context.* Glenview, IL: Scott, Foresman, 1973. 142p.

A book of readings, designed for a college level course in history, that offers material under the following topics: human attitudes toward the earth; roots of environmental decay; nature abused; the blight of urban life; and toward an ethic of environment. Not indexed.

5-136 Detwyler, Thomas R. *Man's Impact on Environment.* New York: McGraw-Hill, 1971. 731p.

A college text by a geographer, which presents a variety of readings by well-known authors. The book is arranged in ten sections: introduction, some basic causes, man's impact on atmosphere and climate, man's impact on the waters, man's impact on land and soils, the spread of organisms by man, destruction of vegetation, destruction and extinction of animals, man as maker of new plants and animals, and trends and prospects. A glossary and bibliography are included.

5-137 Dice, Lee Raymond. *National Communities.* Ann Arbor: University of Michigan Press, 1952. 547p.

According to the author, the purpose of this book is "to describe the more important methods and concepts of one division of biology, that dealing with natural communities." Selected references appear at the end of each chapter, and there is an extensive bibliography of literature cited. Indexed by subject.

5-138 Dickinson, Robert Eric. *Regional Ecology: The Study of Man's Environment*. New York: Wiley, 1970. 199p.

A book by a British geographer, now a resident of the United States, which utilizes some of the concepts of ecology to present a geographical analysis of regionalism in all its European and North American manifestations. More important as a contribution to geography than to ecology, this book shows how other disciplines have been influenced by ecology's emergence as a separate area of scholarship. Indexed.

5-139 Disch, Robert. *The Ecological Conscience: Values for Survival*. Englewood Cliffs, NJ: Prentice-Hall, 1970. 206p.

A collection of readings arranged in sections on the ecological conscience, the impact of ecological values, and the relationship between ecology and social institutions. Well-known writers are reprinted in this anthology, which contains a final chapter on the metaphysics of ecology. A bibliography of suggested readings concludes the volume.

5-140 Dodd, Edward. *Careers for the '70s: Conservation*. New York: Crowell-Collier, 1971. 181p.

The author, creator of the *Mark Trail* comic strip, presents the requirements, advantages, and disadvantages of conservation careers in a straightforward way. This book is limited to conservation rather than the full range of environmental problems. It includes a discussion of salaries with a chart of civil service pay at different levels. A chapter on summer conservation jobs includes addresses where applications should be sent. Another chapter tells where to write for career information. A brief bibliography is appended, along with a directory of universities and the programs they offer. Illustrated. Indexed. Grades 7+.

5-141 Dolan, Edwin G. TANSTAAFL: *The Economic Strategy for Environmental Crisis*. New York: Holt, 1971. 115p.

This book attempts to introduce the reader to the fundamentals of economics, and to supply an ecological critique of the discipline. Chapters on pollution and the price system, environmental problems and economic development, and preserving wilderness. Not indexed.

5-142 Dorfman, Robert and Dorfman, Nancy. *Economics of the*

Environment: Selected Readings. New York: Norton, 1972. 426p.

The editors, both professors of economics, have chosen twenty-six essays which they feel "survey the key facets of the problem from an economic point of view." Contributors include Allen Kneese, Milton Friedman, Barry Commoner, and John Kenneth Galbraith. The essays are arranged in five sections, on the nature of the problem, concepts and methods of economic analysis, pros and cons of policies for environmental protection, reasons for rapid environmental deterioration, and methods and difficulties in making quantitative assessments of environmental damage or costs. Charts and graphs. List of further readings included at end of volume. Not indexed.

5-143 Dorst, Jean. *Before Nature Dies.* Translated by Constance D. Sherman. Boston: Houghton Mifflin, 1970. 352p.

Amply illustrated with charts and photographs, this book deals with human beings and their relationship to nature. The author is the curator of the Division of Mammals and Birds of the National Museum of Natural History in Paris. Abuses throughout the world are cataloged. Indexed, with an extensive bibliography.

5-144 Douglas, William Orville. *The Three Hundred Year War: A Chronicle of Ecological Disaster.* New York: Random House, 1972. 215p.

A catalog of the environmental abuses human beings are responsible for in North America, and throughout the world. Chapters on air, water, radiation, pesticides, garbage, noise, and a number of other topics. A list of federal environmental agencies and federal legislative committees completes the volume. Indexed.

5-145 Dubos, René Jules. *A God Within.* New York: Scribner, 1972. 325p.

This "optimistic philosophy" about the condition and potential of mankind emphasizes the "distinctive genius of each place and each person." Written by a world-famous microbiologist and author, the book is a rationale for positive thinking about earth's destiny. Reference notes are gathered at the end of the volume. Brief selected bibliography. Indexed.

5-146 Dubos, René Jules. *So Human an Animal*. New York: Scribner, 1968. 267p.

An account of man's interaction with his environment by a noted microbiologist. Written for the layman, this book on human ecology is based on "the faith that it is possible to deal scientifically with the living experience of man," and attempts to explain how we are shaped by our surroundings and events. Optimistically, the author proposes a "science of humanity" as the key to humanity's continued existence on earth. Popular and not technical. Indexed.

5-147 Dugan, Patrick R. *Biochemical Ecology of Water Pollution*. New York: Plenum, 1972. 159p.

This is a technical examination of water pollution which proceeds by "interrelating gross examinations with biological reactions occurring at the cellular and molecular level." The introductory material presented on each major topic is accessible to the general reader, but a background in chemistry would be required to make full use of the text, as many formulas are presented. References are gathered at the end of the volume. Illustrated. Indexed.

5-148 Dwiggins, Don. *Spaceship Earth: A Space Look at Our Troubled Planet*. San Carlos, CA: Golden Gate, 1970. 80p.

This is a sophisticated but clearly written book for young readers, especially children interested in space technology. The author explains the operation and uses of Earth Resources Technology Satellites and the Earth Resources Operational Satellite System. Interweaving an outline of the environmental problems now facing the earth, the author shows how these special satellites can help pinpoint zones of chemical or thermal pollution, crop destruction, and polluted air and water. Black-and-white drawings and photos are used effectively to illustrate the somewhat technical text. Glossary. Briefly indexed. Grades 5-8.

5-149 Dye, Lee. *Blowout at Platform A: The Crisis That Awakened a Nation*. Garden City, NY: Doubleday, 1971. 231p.

A book that recounts the blowout of an oil well in the Santa Barbara channel and the effect which that event had upon the petroleum industry. Indexed.

5-150 Easton, Robert. *Black Tide: The Santa Barbara Oil Spill and Its Consequences.* New York: Delacorte, 1972. 336p.

The story of the eruption of an oil well off Santa Barbara, California, on January 28, 1969, and its effects on the environment and the community. The book, with an introduction by Ross McDonald, carries in its appendix a chronology of the Santa Barbara oil spill, a list of major oil spills, and an extensive bibliography. Indexed, with some photographs.

5-151 *Eco-catastrophe.* By the editors of *Ramparts.* New York: Harper, 1970. 158p.

A collection of eleven articles reprinted from *Ramparts* magazine, this small volume is the anti-establishment statement on the environmental crisis. It "articulates a radical perspective, locating the root of the ecological crisis in the very structure of American society" and calls for "revolutionary reconstruction of the social order" before any real change can occur. Illustrated.

5-152 *Ecology: Crisis & New Vision.* Edited by Richard E. Sherrell. Richmond, VA: Knox, 1971. 159p.

This book is based on the proceedings of a symposium sponsored by the Church Society for College Work and the Department of Higher Education of the National Council of Churches. The essays address the philosophical and theological issues raised by the environmental crisis, the social and economic ethics involved, and the potential of religious imagination in dealing with human values.

5-153 *Ecology in Theory and Practice.* Edited by Jonathan Benthall. New York: Viking, 1973. 367p.

A series of lectures, given at the Institute of Contemporary Arts (London) in 1970–1971, and published in the United Kingdom under the title *Ecology, The Shaping Enquiry.* There are sections on model ecosystems and contemporary problems. Essays on chemical and biological warfare, ecology and the computer, ocean resources, industrial pollution, cities, the influence of technology on the environment, and other topics. Indexed.

5-154 *The Economics of Environmental Quality.* By James C. Hite, Hugh H. Macaulay, James M. Stepp, and Bruce Yandle, Jr.

Washington, DC: American Enterprise Institute for Public Policy Research, 1972. Domestic Affairs Study 5. 113p.

The authors use the term "environmental quality" to refer to the "conditions associated with those resources that have not been assigned to the market for allocation." The short book focuses on air and water quality, and offers reasons why "although environmental quality of a high level is desirable the additional benefits of improved quality may not be worth their additional cost." Not indexed.

5-155 *The Ecosystem Concept in Natural Resource Management.* Edited by George Van Dyne. New York: Academic, 1969. 383p.

This book is based on a symposium held at the annual meeting of the American Society of Range Management in Albuquerque, New Mexico, February 12–15, 1968. The papers included cover range, forest, watershed, fishery, and wildlife resource management. Each paper has its own list of references, many of them extensive. Author and subject indexes.

5-156 Edel, Matthew. *Economies and the Environment.* Englewood Cliffs, NJ: Prentice-Hall, 1973. 162p.

Organized in seven sections: economics and ecology; population resource balance; institutions and food supply; growth, pollution, and social cost; environmental fine tuning; the social organization of internal combustion; and economic systems. Each chapter contains recommendations for further reading. Indexed.

5-157 Edwards, Clive A. *CRC Persistent Pesticides in the Environment.* Cleveland: CRC Press, 1970. 78p.

A short monograph that details the occurrence and persistence of pesticides in the physical environment and in biota. There is a chapter on the control of pesticides, and an extensive list of references is cited in the text. Illustrated with charts. This book appeared as part of an article in the *CRC Critical Reviews in Environmental Control.* Not indexed.

5-158 Ehrenfeld, David W. *Conserving Life on Earth.* New York: Oxford, 1972. 360p.

A revised edition of the author's *Biological Conservation*, intended as a trade book for the general public. It contains a chapter on the blue whale and is amply illustrated. Annotated bibliography. Indexed.

5-159 Ehrlich, Paul R. *The Population Bomb*. New York: Ballantine, 1968. 223p.

An important work, now almost a contemporary classic, that argues persuasively for population control. There is a chapter on what individuals can do, and an appendix carries samples of letters which can be written to politicians and religious leaders. A selected bibliography is included. Indexed.

5-160 Ehrlich, Paul R. and Ehrlich, Anne H. *Population, Resources, Environment: Issues in Human Ecology*. 2d ed. San Francisco: Freeman, 1972. 383p.

This is a very comprehensive and lucid textbook on population and resources by authors well known in the field. Among the many topics covered are demographic statistics, limits of the earth, food production, environmental threats, endangered ecosystems, optimum population, family planning, birth control, and an international perspective on population problems. The final section offers recommendations for action. Each chapter includes a bibliography, and there is a general bibliography too. Much statistical material appended. Indexed.

5-161 Ehrlich, Paul R.; Ehrlich, Anne H.; and Holdren, John P. *Human Ecology: Problems and Solutions*. San Francisco: Freeman, 1973. 302p.

Less detailed than Ehrlich's *Population, Resources, Environment*. This survey focuses on the biological and physical aspects of environmental problems and proposes solutions. An appendix contains 1972 world population data. Indexed. Chapters include bibliographies for further reading.

5-162 Ehrlich, Paul R. and Harriman, Richard L. *How To Be a Survivor*. New York: Ballantine, 1971. 208p.

Using the metaphor of a spaceship, the authors deal with overpopulation, overdevelopment, and government control in relation to the environment. The concept of ODC and UDC (overdeveloped and underdeveloped countries) is central to the authors' analysis and to their final recommendations. A large appendix contains recommendations for action, a bibliography, and a reprint of a proposed "Constitution for a United Republics of

America" that first appeared in *The Center Magazine*, published by The Center for the Study of Democratic Institutions.

5-163 Elliot, Sarah M. *Our Dirty Air*. New York: Messner, 1971. 64p.

A simple but serious book about air pollution. The text begins with a description of the 1970 crisis in Japan and the United States. Other topics include causes of air pollution, air pollution and weather, the effect of pollution on living things, and what can be done about air pollution. Illustrated with black-and-white photographs and line drawings. Indexed. Grades 3-5.

5-164 Elliot, Sarah M. *Our Dirty Water*. New York: Messner, 1973. 64p.

This book follows the format of the author's earlier work, *Our Dirty Air* (5-163). After a discussion of the importance of water, the author deals with pollution, including who pollutes, what is being done about it, and what the reader can do. Illustrated with black-and-white photographs and maps drawn by the author. Indexed. Grades 3-5.

5-165 Elton, Charles S. *Ecology of Invasions by Animals & Plants*. London: Methuen, 1958. 181p.

The book brings together faunal history, ecology, and conservation. Chapters on the invasions of continents, changes in the sea, and the fate of remote islands. Two chapters on conservation. Illustrated with maps and photographs. Extensive references. Indexed.

5-166 Emlen, J. Merritt. *Ecology: An Evolutionary Approach*. Reading, MA: Addison-Wesley, 1973. 493p.

An advanced college text, treating ecology from the standpoint of evolutionary theory. Elementary college-level mathematics is essential to understanding this work. Chapters on mechanisms of evolution, ecology of individuals, ecology of populations, and ecology of communities. Indexed.

5-167 Emmel, Thomas C. *An Introduction to Ecology and Population Biology*. New York: Norton, 1973. 196p.

A concise elementary presentation. According to the author, it was written for those "who want a good basic understanding of their environment, the need for preserving it, and the value of the natural populations that depend upon it." Well illustrated, with a bibliography of recommended readings. A glossary is provided. Indexed.

5-168 *Energy and Power.* San Francisco: Freeman, 1971. 144p.

The text of this book originally appeared in the September 1971 issue of *Scientific American*. The several chapters deal with energy flow in the biosphere and in primitive, agricultural, and industrial societies; the conservation of energy; and energy and information. A final article discusses decision-making in the production of power. Brief biographies of the contributors and a selected bibliography are included. Indexed. Illustrated, partly in color.

5-169 *Energy, Economic Growth, and the Environment; papers presented at a forum conducted by Resources for the Future, Inc., in Washington, DC, April 20–21, 1971.* Edited by Sam H. Schurr. Baltimore: Johns Hopkins University Press, 1972. 232p.

The nine papers included here range in length from 10 to 30 pages and deal with economic growth and increased energy consumption, the impact of both on the environment, and issues of public policy. Indexed.

5-170 *Environment: A National Mission for the Seventies.* By the editors of *Fortune.* New York: Harper, 1970. 220p.

This material first appeared in the October 1969 and February 1970 issues of *Fortune.* Thirteen chapters, among them "What Business Thinks About its Environment," "Conservationists at the Barricades," and the "Economics of Environmental Quality." Not indexed.

5-171 *The Environment Crisis: A Basic Overview of the Problem of Pollution.* By Daniel M. Rohrer, and others. Skokie, IL: National Textbook, 1970. 346p.

This textbook contains chapters on the technology of pollution; the economics of pollution; methods of control; existing programs; federal, state and local governments' roles in the control of pollution; and industry's role in pollution abatement. Illustrated with charts and graphs. There is an extensive section of authors' notes, and a lengthy and useful bibliography. Not indexed.

5-172 *The Environmental Crisis: Agenda for Survival.* Edited by Harold W. Helfrich, Jr. New Haven: Yale University Press, 1971. 234p.

The second volume (see below) derived from the two-year Yale School of Forestry Symposium, Issues in the Environmental Crisis, sponsored by the Ford Foundation. Essays by Stewart Udall, Paul McCloskey, Charles Luce, Harry Caudill, and others. A solid collection. Not indexed.

5-173 *The Environmental Crisis: Man's Struggle To Live With Himself.* Edited by Harold W. Helfrich, Jr. New Haven: Yale University Press, 1970. 187p.

A book of lectures given at the 1968–1969 symposium at the Yale University School of Forestry (see above). Essays by LaMont C. Cole, Paul Ehrlich, Joseph L. Sax, Kenneth Boulding and others. Not indexed.

5-174 *Environmental Mercury Contamination.* Edited by Rolf Hartung and Bertram D. Dinman. Ann Arbor, MI: Ann Arbor Science, 1972. 349p.

This volume is based on the International Conference on Environmental Mercury Contamination held in Ann Arbor in 1970. The scholarly papers are organized into sections on the occurrence of mercury in the environment and in human beings, methods of analysis, environmental dynamics of mercury, and biological effects of mercury compounds. Illustrated. Indexed.

5-175 *Environmental Problems: Pesticides, Thermal Pollution and Environmental Synergisms.* Edited by Billy Ray Wilson. Philadelphia: Lippincott, 1968. 183p.

This book is the result of a conference entitled "Graduate Opportunities in Environmental Health Sciences," held at Rutgers in April 1967. The eleven papers cover both general problems of definition and scope and specific problems, such as the use of pesticides and the effects of thermal pollution. A list of conference participants precedes the table of contents and a list of colleges represented at the conference precedes the index.

5-176 *Environmental Quality and Social Behavior: Strategies for Research; report on a study conference on research strategies in the social and behavioral sciences on environmental problems and policies.* Washington, DC: National Academy of Sciences, 1973. 86p.

This book "presents the rationale for intensified environmental research in the social and behavioral sciences and outlines one of several approaches that could be taken to constructing strategies for its pursuit." An Environmental Quality Research Corporation is proposed as a nonprofit, government-funded, and privately controlled corporation to undertake needed research.

5-177 Fabricant, Neil and Hallman, Robert Marshall. *Toward a Rational Power Policy: Energy, Politics, and Pollution; a report by the Environmental Protection Administration of the City of New York.* New York: Braziller, 1971. 292p.

A report on the health and environmental problems associated with the generation of electricity in New York state. This book is of general interest for the information it contains; useful bibliography.

5-178 Fadiman, Clifton and White, Jean, eds. *Ecocide—And Thoughts Toward Survival.* Santa Barbara, CA: Center for the Study of Democratic Institutions, 1971. 202p.

Many of these essays were written for a conference devoted to ecology held by the Center for the Study of Democratic Institutions in 1970. The essays are grouped under two topics: the environmental crisis and thoughts toward survival. Some of the material has previously been published in *The Center Magazine.* Not indexed.

5-179 Faith, William Lawrence and Atkinson, Arthur A., Jr. 2d ed. New York: Wiley–Interscience, 1972. 393p.

A multidisciplinary treatment of the air quality management field. It includes chapters on meteorology, smoke, gases, odors, photochemical air pollution, and social origins of air pollution. Appended is a list of conversion factors for common air pollution measurement. Indexed.

5-180 Falk, Richard A. *This Endangered Planet: Prospects and Proposals for Human Survival.* New York: Random House, 1971. 495p.

An examination of the possible scenarios for international political solutions to the problems of the earth's environment. A new "world order" is proposed. A lengthy appendix contains chapter footnotes, a table of world military expenditures, the 1969 population data for 137 countries, a bibliography for further reading, and a short list of organizations concerned with human survival. Useful as a book that demonstrates how ecology has affected political science. Indexed.

5-181 Fallows, James M. *The Water Lords: Ralph Nader's Study Group Report on Industry and Environmental Crisis in Savannah, Georgia.* New York: Grossman, 1971. 294p.

The author and his several colleagues, all members of the Savannah River Project, spent a summer making an intensive study of the abuse of the Savannah River. A historical sketch of the development of Savannah and use of the river is followed by a careful look at the political and economic structure of the city. The investigation focuses on the sewage treatment plant which the city was supposed to complete in 1968 and then planned to finish in 1973, and on the large pulp and paper mills which pollute the river. As is typical of the Nader reports, this study is carefully documented.

5-182 Fanning, Odom. *Opportunities in Environmental Careers.* Louisville, KY: Vocational Guidance, 1971. 271p.

This book attempts to catalog present needs in the environmental sciences and project personnel needs for the future. The author covers two-year environmental technology curricula and four-year and postgraduate environmental sciences curricula. He then details the nature of and requirements for careers in ecology, earth sciences, resources and recreation, environmental design, and environmental protection. Each chapter concludes with a list of professional organizations that offer literature on education and careers. Appended are a bibliography and lists of relevant periodicals, activist organizations, key government agencies, institutions offering training in environmental technology, and one hundred university environmental science centers. Indexed.

5-183 Farb, Peter. *Ecology.* New York: Time-Life Books, 1970. 192p.

Chapters on the rise and fall of populations and why living things are where they are. Amply illustrated, the book carries a short bibliography and presents the major biomes of the world as they would be if they were undisturbed by man. Indexed. First published in 1963.

5-184 Fisher, James; Simon, Noel; and Vincent, Jack. *Wildlife in Danger.* New York: Viking, 1969. 368p.

This extensive account of the animals, birds, and plants of the world threatened with extinction is based on facts collected by the Survival Service Commission of the International Union for Conservation of Nature and Natural Resources. There are separate chapters on mammals, birds, reptiles, amphibians, fishes, and plants, with many color and black-and-white illustrations throughout. Some 150 of the illustrations were commissioned for this book by wildlife artists from around the world. Each species is discussed in one or two pages of text. The volume also includes an index of common and scientific names, and maps on the endpapers that show world distribution of endangered mammals and birds.

5-185 *The Fitness of Man's Environment.* Washington, DC: Smithsonian Institution Press, 1968. Smithsonian Annual II. 250p.

This volume brings together a group of essays by individuals from varied backgrounds but with a common concern about the quality of the environment. Essays by Paul Goodman, Robert M. Adams, Philip Johnson, Edward T. Hall, and others. The city and our relationship to urban living is a major focus of the essays presented. The book results from a symposium held February 16–18, 1967. Not indexed.

5-186 Flawn, Peter T. *Environmental Geology: Conservation, Land-Use Planning, and Resource Management.* New York: Harper, 1970. 313p.

This college text is one volume in Harper's Geoscience Series. The author, director of the Texas Bureau of Economic Geology, defines environmental geology as a branch of ecology that "deals with relationships between man and his geological habitat." Beginning with sections on the earth processes and properties of rocks and soils, the volume also deals with the geological consequences of industrialization, conservation and management, and the application of geologic data. Illustrated with tables and diagrams. Appended materials include a building code, classification of rocks, and a substantial glossary. Also included loose is a full-color geologic quadrangle map of part of Austin, Texas, illustrating the final section of the text which uses Austin as a case study. Indexed.

5-187 Forstner, Lorne J. and Todd, John H., eds. *The Everlasting Universe: Readings on the Ecological Revolution.* Lexington, MA: Heath, 1971. 370p.

A collection of readings on ecology arranged in the following sections: ecology—a definition; nature—background for the crisis; the crisis; toward a future. A great deal of this reprinted material is popular, rather than technical, in nature. Essays by Norman Mailer, Saul Bellow, D. H. Lawrence, John Fischer, Gary Snyder, John Updike, and other literary figures. Not indexed.

5-188 Foss, Phillip O. *Politics and Ecology.* Belmont, CA: Duxbury, 1972. 298p.

A collection of readings intended for the layman, this book concentrates on the political aspects of environmental problems. The editor has chosen essays by well-known figures such as Edmund Muskie, Aldous Huxley, William O. Douglas, Rachel Carson, and Gaylord Nelson, and by many lesser-known authors, scientists, and political scientists. Essays are grouped into sections on definition of the problems, water pollution, air pollution, marine pollution, population, and the politics of survival. Selected additional readings. Not indexed.

5-189 Fraser, Dean. *The People Problem: What You Should Know About Growing Population and Vanishing Resources.* Bloomington: Indiana University Press, 1971. 248p.

A book on human ecology that focuses on the population explosion and its implications for human survival. A final section deals with population control. Indexed, with a brief bibliography of recommended readings. The second part of the book contains an extensive section on factors which limit population growth, most of which are environmental.

5-190 Friedmann, Wolfgang Gaston. *The Future of the Oceans.* New York: Braziller, 1971. 132p.

A book by a professor of international law on managing the oceans and the oceanbed ecologically. According to the book, "navigation, as well as the exploitation of the living and mineral resources of the sea, must be the subject of planning and regulation for the common benefit of mankind." Indexed.

5-191 Friends of the Earth. *The Stockholm Conference—Only One Earth; an introduction to the politics of survival.* London: Earth Island, 1972. 172p.

An attractively illustrated book that interweaves the proposals for action presented at the Stockholm Conference with a history of the earth, written from the standpoint of ecological concern. Not indexed.

5-192 Frome, Michael. *Battle for the Wilderness.* New York: Praeger, 1974. 246p.

The concept of wilderness in the United States is explored historically. The second major section of the book deals with steps that have been taken to preserve wilderness, culminating in the Wilderness Act of 1964. Indexed.

5-193 Fuller, Richard Buckminster. *Operating Manual for Space-ship Earth.* New York: Simon & Schuster, 1969. 143p.

Architect and research professor Buckminster Fuller offers a speculative analysis of human intellectual evolution and a positive prognosis for survival on "spaceship earth." Although there is little of direct environmental interest here, this is an essential title for an environmental collection.

5-194 Gabrielson, Ira N. *Wildlife Conservation.* 2d ed. New York: Macmillan, 1959. 244p.

The author tries to put into simple language the basic facts of wildlife conservation. The book is divided into two sections: The first seven chapters show the interdependence of conservation programs while the remaining chapters deal with problems of certain groups of wildlife. There are chapters on race and vanishing species, wildlife refuges, and the "obstacles" to conservation. Illustrated. Indexed.

5-195 Garvey, Gerald. *Energy, Ecology, Economy.* New York: Norton, 1972. 235p.

An attempt to place our national energy policy into a coherent framework of relevant environmental concepts. The model of an "ecological perspective" in fuel use is advanced in chapter 3, while an appendix offers a formal analysis of the concepts presented. Air pollution, nuclear power, water quality, and the environmental costs of coal are treated. Indexed, with a glossary.

5-196 Giddings, John Calvin. *Chemistry, Man, and Environmental Change: An Integrated Approach.* San Francisco: Canfield, 1973. 472p.

This textbook on chemistry and the environment is intended

for "nonscience students who desire an introduction to the science of environment without undue burden of scientific detail." The first five chapters establish a background in chemistry and the last six explore major environmental problems such as air pollution, water pollution, and nuclear energy. Each chapter includes exercises and a glossary, and the last six chapters also contain lists of additional readings. Illustrated. Indexed.

5-197 Giddings, John Calvin and Monroe, Manus B., eds. *Our Chemical Environment.* San Francisco: Canfield, 1972. 367p.

This collection of reprinted articles focuses on the chemical nature of the environment and man. It covers air pollution, water abuse, contamination by metals, pesticides, food additives, solid wastes, energy, and nuclear power. Not indexed.

5-198 Gill, Don and Bonnett, Penelope. *Nature in the Urban Landscape: A Study of City Ecosystems.* Baltimore: York, 1973. 209p.

A study of wildlife in urban areas, and the environmental factors that can affect the survival of natural communities of wildlife. Chapters on London and Los Angeles. There is, in addition, material on planning for wildlife in the city and managing urban wildlife habitats. Illustrated with photographs and charts. Lengthy bibliography of research in urban wildlife ecology, a glossary, and tables of scientific and common names. Indexed.

5-199 Gilliam, Harold. *For Better or For Worse: The Ecology of an Urban Area.* San Francisco: Chronicle, 1972. 183p.

A history and survey of the environmental crisis affecting the San Francisco Bay area. Much of the material first appeared in the *San Francisco Chronicle*'s Sunday magazine, where the author writes a regular conservation column. Illustrated with photographs. Not indexed.

5-200 Gofman, John W. and Tamplin, Arthur R. *Poisoned Power: The Case Against Nuclear Power Plants.* Emmaus, PA: Rodale, 1971. 368p.

The authors of this book were asked in 1963 by the Atomic Energy Commission to undertake a series of long-range studies on potential dangers of peaceful uses of the atom. The Commission was unhappy with the results of the study. This book is a strong statement on the serious dangers at existing nuclear power plants and the lack of enforced safety standards. Foreword by Senator Mike Gravel. Appended material includes a list of often-asked questions with the authors' answers, a list of moratorium activists, a list of the names and addresses of members of the Atomic Safety and Licensing Board Panel, and a list of U.S. Central Station Nuclear Power Projects. Not indexed.

5-201 Goldman, Charles R.; McEvoy, James, III; and Richerson, Peter J., eds. *Environmental Quality and Water Development: Summary and Recommendations.* San Francisco: Freeman, 1973. 510p.

This volume is based on a study undertaken by the editors in 1970 for the National Water Commission on "the cause of conflict between environmental quality and the development of the nation's water resources." The study emphasizes attitudes and values as they have affected political and economic views of how water resources should be used. Chapter 1 contains the recommendations offered by the authors. The volume also includes essays and articles by people consulted during the study on the history of water development, psychological responses to the environment, demographic effects of water development, water development and urban recreation, and other topics. Brief list of references attached to most chapters. Indexed.

5-202 Goldman, Marshall I. *Controlling Pollution: The Economics of a Cleaner America.* Englewood Cliffs, NJ: Prentice-Hall, 1967. 175p.

A collection of readings, assembled by an economist, on the cost and problems of ending pollution. A lengthy essay introduces the book and one major section is devoted to the economic analysis of pollution abatement. Essays by economists, journalists, and others interested in the problem. (See 5-203.)

5-203 Goldman, Marshall I. *Ecology and Economics: Controlling Pollution in the 70s.* Englewood Cliffs, NJ: Prentice-Hall, 1972. 234p.

This collection of articles is an updated version of *Controlling Pollution: The Economics of a Cleaner America,* published in 1967 (5-203). The reprinted articles detail what is being done to cope with pollution in the United States, Japan, Germany, and the Soviet Union. Using case studies, several authors show how conditions have been improved, while others examine instances of neglect. Extensive cost estimate figures show the price of cleaning up pollution. Not indexed.

5-204 Goldstein, Jerome. *How to Manage Your Company Ecologically*. Emmaus, PA: Rodale, 1971. 119p.

This is a straightforward, logical book written by the executive editor of Rodale Press. He begins by admitting the sins of his own publishing house, such as the mailing of thousands of circulars (junk mail) which become refuse. His approach to managing a company ecologically is not naive; he knows that a business has to be profitable and seeks ways to make conservation economical. He deals with recycling, using a company's purchasing power to advantage, the deception of empty rhetoric in advertising, the need for electric vehicles, and the need for laws that make ecological decisions profitable. He points out, for example, that the only way to have decent mass transportation is to allow General Motors to produce and profit from the new system. Illustrated with cartoons. Suggested reading list included. Not indexed.

5-205 Goodman, Gordon T.; Edwards, R. W.; and Lambert, J. M., eds. *Ecology and the Industrial Society; A Symposium of the British Ecological Society, Swansea, 13–16 April 1964*. New York: Wiley, 1965. The British Ecological Society. Symposium No. 5. 395p.

Technical contributions on a number of subjects, including water pollution, the ecology of marine fouling, the ecology of pests in shops and homes, and the oxygen balance of fresh water streams. Indexed by subject and author.

5-206 Gorden, Morton and Gorden, Marsha. *Environmental Management: Science and Politics*. Boston: Allyn & Bacon, 1972. 548p.

Described as a "second generation" book, this volume "seeks to identify the kinds of information scientists can produce to best integrate their professional concerns into the political processes of environmental management." Most of the reprinted articles are technical, but some are not. There are original interviews with Stanley M. Greenfield and Russell Train. Indexed.

5-207 Grad, Frank P.; Rathjens, George W.; and Rosenthal, Albert J. *Environmental Control: Priorities, Policies, and the Law*. New York: Columbia University Press, 1971. 311p.

This book is the result of a study prepared by the Legislative

Drafting Research Fund of Columbia University. The papers included were, in their original form, presented at a Symposium on Federal Environmental Policy sponsored by the Association of the Bar of the City of New York in May 1970. The papers are organized into three sections (each author is responsible for a section). Part 1 is entitled "National Environmental Policy: Goals and Priorities"; part 2, "Intergovernmental Aspects of Environmental Controls"; and part 3, "Federal Power to Preserve the Environment: Enforcement and Control Techniques." Part 1 includes a bibliography, while parts 2 and 3 conclude with lengthy notes. Not indexed.

5-208 Graham, Frank, Jr. *Since Silent Spring.* Boston: Houghton Mifflin, 1970. 333p.

This book deals with the controversy surrounding the use of pesticides from the publication of Rachel Carson's *Silent Spring* in 1962 until 1970. The author documents the furor surrounding *Silent Spring*, attempts within industry and government to discredit Rachel Carson, and the evolving fight to control dangerous chemicals in the environment. Outlining the contradictory policies of government, Graham indicates some actions open to the citizenry. Appendixes include an article on safer pesticides for home and garden use and a discussion of federal registration requirements for pesticide products. Indexed.

5-209 Grayson, Melvin J. and Shepard, Thomas R., Jr. *The Disaster Lobby: Prophets of Ecological Doom and Other Absurdities.* Chicago: Follett, 1973. 294p.

An anti-ecological treatise on what the authors call "the disaster lobby." The environmental issues and concerns which have occupied the 1960s and 1970s are discussed; according to the authors, "the age of the disaster lobby has run out of steam." Indexed. Useful as an example of popular criticism of the proponents of environmental concern.

5-210 Gregor, Arthur S. *Man's Mark on the Land: The Changing Environment, from the Stone Age to the Age of Smog, Sewage, and Tar on Your Feet.* New York: Scribner, 1974. 120p.

After documenting the ascent of the human race, the author discusses the dangers and problems we have inflicted on ourselves. The book includes a "Dictionary for the Environment" and a list of materials for further reading. Illustrated with black-and-white photographs. Indexed. Grades 5-8.

5-211 Grossman, Shelly and Grossman, Mary Louise. *The How and Why Wonder Book of Ecology.* New York: Grosset, 1971. 47p.

This book follows the format of the How and Why series, with large type and many photographs. The book discusses food chains, habitats, and biomes, and then treats basic ecosystems, such as grasslands, deserts, mountains, and coastlines. A final section deals with the problems of using DDT and the alternatives to its use. Not indexed. Grades 2-5.

5-212 *Growing Against Ourselves: The Energy-Environment Tangle; problems, policies, and approaches.* Edited by S. L. Kwee and J. S. R. Mullender. Lexington, MA: Lexington Books, 1972. 252p.

This scholarly monograph is number 6 of the Publications of the John F. Kennedy Institute, Center for International Studies, Tilburg, The Netherlands. It comprises most of the papers prepared for a colloquium held by the Institute and the Future Shape of Technology Foundation in The Hague in 1971. The papers are organized into three sections: electric energy and environment—identifying the problems; technical problems and national energy policies; and future strategies and approaches. Indexed.

5-213 Guitar, Mary Anne. *Property Power: How to Keep the Bulldozer, the Power Line, and the Highwaymen Away from Your Door.* Garden City, NY: Doubleday, 1972. 322p.

A history of attempts by individuals and citizen groups to prevent unscrupulous land development. Problems and successes are described. Brief bibliography. Indexed.

5-214 Haefele, Edwin T. *Representative Government and Environmental Management.* Baltimore: Johns Hopkins University Press (for Resources for the Future), 1973. 188p.

Explores the concept of environmental management. Most of the material has been published in some form elsewhere. Chapters are devoted to governance of common-property resources; residuals management, metropolitan governance, and the optimal jurisdiction; and environmental quality as a problem of social choice. Most of the material is technical in nature. Not indexed.

5-215 Hahn, James and Hahn, Lynn. *Recycling: Re-using Our World's Solid Wastes.* New York: Watts, 1973. 66p.

In a well-organized text, the authors discuss the overwhelming amounts of solid waste that accumulate in the world. They present ways in which individuals can help by collecting and sorting re-usable materials. A glossary is appended. Indexed. Grades 4-7.

5-216 Hall, Gus. *Ecology: Can We Survive Under Capitalism?* New York: International Publishers, 1972. 94p.

Ecology as the subject matter for political action, by the current General Secretary of the Communist Party, U.S.A. Socialism is represented as the answer to worldwide problems of pollution and the only solution for the industrial countries in the West. Not indexed.

5-217 Hamilton, Michael Pollock. *This Little Planet.* New York: Scribner, 1970. 241p.

Essays by Paul Sears, Rev. William G. Pollard, Ivan Bennett, Jr., Roger Shinn, Clarence Glacken, and Conrad Bonifazi on pollution, scarcity, and conservation. Senator Edmund Muskie provides an introduction. Not indexed.

5-218 Hammond, Allen L.; Metz, William D.; and Maugh, Thomas H., II. *Energy and the Future.* Washington, DC: American Association for the Advancement of Science, 1973. 184p.

A book which "seeks to discover and to assess the technologies and research developments that will be the basis for future energy policies." Major sections are devoted to energy from fossil fuels, nuclear energy, alternative energy sources, energy transmission, and energy conservation. A final chapter deals with research priorities and energy policy. The book contains a glossary and a bibliography of chapter notes. Indexed.

5-219 Hardin, Garrett. *Exploring New Ethics for Survival: The Voyage of the Spaceship "Beagle."* New York: Viking, 1972. 273p.

Biologist Garrett Hardin is primarily concerned with the need for population control in a world of limited resources. He is a sophisticated writer, beginning each of the three sections of his book with a continuing science fiction parable which helps the reader perceive the scope and purpose of the remainder of the text. The book "unifies problems of pollution control, population control, and the wise use of natural resources." Notes and references are gathered at the end of the volume. The original essay by the author, "The Tragedy of the Commons," which was the nucleus for this book, is appended. Indexed.

5-220 Harmer, Ruth Mulvey. *Unfit for Human Consumption*. Englewood Cliffs, NJ: Prentice-Hall, 1971. 374p.

A book about pesticides and the part played by federal agencies, especially the Department of Agriculture, in encouraging their use. DDT and the Shell "No Pest" strip are among the topics covered. Indexed.

5-221 Harris, Larry. *Twilight of the Animal Kingdom: The Endangered Species*. Los Angeles: Ward Ritchie, 1972. 48p.

The author presents twenty-two endangered species, describing each in a clear, one-page statement. The scientific name, order, and family are given, as well as weight, distribution of remaining examples, reason for potential extinction, and laws now protecting the species. The author is also an artist and depicts each animal in a full-page color illustration. Grades 3-8.

5-222 Harrison, C. William. *A Walk Through the Marsh*. Chicago: Reilly & Lee, 1972. 32p.

Clearly illustrated with photographs, this is a book about the ecology of a marsh, with information on predators and migratory patterns, and a brief discussion of pollution and conservation. Not indexed. Grades 3-5.

5-223 Harrison, Gordon A. *Earthkeeping: The War With Nature and a Proposal for Peace*. Boston: Houghton Mifflin, 1971. 276p.

Written by the director of the Ford Foundation's Resources and Environment program, this book provides a general argument for environmental concern on the part of the average citizen. Separate and extensive notes follow the text. Indexed.

5-224 Harte, John and Socolow, Robert H. *Patient Earth*. New York: Holt, 1971. 364p.

Ten case studies on environmental problems are the nucleus of this book. Each reflects positive accomplishments and is presented by individuals who were directly involved with these "environmental successes." Additional sections include "Lessons from the Past" and "Constraints on Growth: The Scientific Argument." Bibliography. Indexed.

5-225 Haskell, Elizabeth H. and Price, Victoria S. *State Environmental Management*. New York: Praeger, 1973. 283p.

A study of environmental regulation and management in Illinois, Minnesota, Washington, Wisconsin, New York, Vermont, Maine, Maryland, and Michigan. Bibliography. Not indexed.

5-226 Hawkins, Donald E. and Vinton, Dennis A. *The Environmental Classroom*. Englewood Cliffs, NJ: Prentice-Hall, 1973. 374p.

A presentation of educational theories based upon learning by interacting with the environment. Part 2 of the book "offers the ideas of some of the people who have . . . worked on the problem of reconnecting human beings to their environment." Indexed.

5-227 Henkin, Harmon; Merta, Martin; and Staples, James, eds. *The Environment, the Establishment and the Law*. Boston: Houghton Mifflin, 1971. 223p.

The proceedings of a hearing before the Wisconsin Department of Natural Resources on whether DDT was polluting the waters of the state. There is a wealth of expert testimony on the effects of DDT in the environment. Amply illustrated with an appendix that contains a model pesticide law and a glossary of terms.

5-228 Herber, Lewis. *Crisis in Our Cities*. Englewood Cliffs, NJ: Prentice-Hall, 1965. 239p.

A solid look at the ecological problems confronting our cities from the standpoint of public health. Mental illness is treated, as well as other diseases that can be traced to the tensions of city life. Indexed.

5-229 Hickel, Walter J. *Who Owns America?* Englewood Cliffs, NJ: Prentice-Hall, 1971. 577p.

A personal account of former Secretary of the Interior Hickel's tenure in office and the environmental crises he confronted. Interesting insights into the politics of the environment as well as broader issues. Includes Hickel's involvement in the aftermath of the Santa Barbara oil spill.

5-230 Hill, Gladwin. *Madman in a Lifeboat: Issues of the Environmental Crisis.* New York: Day, 1973. 118p.

The purpose of this book, according to the author, is to "spotlight major issues on which reform in the principal environmental problem area hinges." The book deals with land, air, and water pollution and solid waste disposal. Useful for the younger reader. Indexed, with a short bibliography of further readings.

5-231 Hilton, Suzanne. *How Do They Get Rid of It?* Philadelphia: Westminster, 1970. 117p.

The author discusses the problems people have in getting rid of what they have created when they no longer want it. The book focuses on recycling as a creative and productive way to handle waste, with separate chapters on recycling automobiles, planes and trains, ships, buildings, containers, wood, paper and books, trash, factory wastes, garbage, sewage, and radioactive waste. Bibliography included. Illustrated with black-and-white photographs. Indexed. Grades 6-8.

5-232 Holdren, John P. and Herrera, Philip. *Energy: A Crisis in Power.* San Francisco: Sierra Club, 1971. 252p.

The authors examine the problems of energy, electric power, and the environment. Part 1, "Energy: Resources and Consumption," is a scientist's overview of the problem, prepared by Holdren. It covers trends in energy use, fossil fuels, hydroelectric power, nuclear fission, thermal pollution, and future energy supplies. Part 2, "Power: Conflicts and Resolutions," was prepared by Herrera, a journalist. It consists of several case studies of controversies over the siting and operation of power plants. Not indexed.

5-233 Holdren, John P. and Ehrlich, Paul R., eds. *Global Ecology: Readings Toward a Rational Strategy for Man.* New York: Harcourt, 1971. 295p.

A collection of essays which emphasizes "the unified nature of the problems in population, resources, and environment." This book draws from many disciplines including the physical sciences, technology, sociology, demography, conservation, politics, and economics. The editors introduce each section and shape the two major themes of the volume—the enormous complexity of the environmental problems before us and the radical changes prerequisite to "the global ecological strategy we need so desperately." Not indexed.

5-234 Hood, Donald W. *Impingement of Man on the Oceans*. New York: Wiley–Interscience, 1971. 738p.

A lengthy volume of technical and scholarly material contributed by specialists, this book is divided into six sections: transport processes and reservoirs, chemical models of the ocean, artifacts of man, man's alteration of coastal environment, models for study of future alterations of the ocean, and implications of man's activities on ocean research. Charts, graphs, and diagrams. Separate author and subject indexes.

5-235 Hopkins, Edward Scott; Bingley, W. McLean; and Schucker, George Wayne. *The Practice of Sanitation in Its Relation to the Environment*. 4th ed. Baltimore: Williams & Wilkins, 1970. 550p.

A textbook on "environmental sanitation," this volume was prepared for health officers, nurses, and sanitary engineers. It includes chapters on solid waste disposal, urban sewage facilities, industrial wastes, stream pollution, and air pollution. Indexed.

5-236 Hottel, Hoyt Clarke and Howard, J. B. *New Energy Technology—Some Facts and Assessments*. Cambridge: MIT Press, 1971. 364p.

Results from a study undertaken by the MIT Environmental Laboratory for Resources for the Future, Inc. The authors, chemical engineers, deal with fossil fuel-to-fuel conversion, nuclear power, and power generation from fossil fuel. Further material is included on the prospects of utilizing solar energy and space heating and cooling. Numerous charts and graphs. Indexed.

5-237 Hudson, Norman. *Soil Conservation*. Ithaca, NY: Cornell University Press, 1971. 320p.

This textbook is quite technical and specialized. The author stresses erosion control as a basic concept in soil conservation and also covers the mechanics of erosion, the physics of rainfall, the principles and design of mechanical protection, land management, wind erosion, erosion research methods, pollution and soil erosion, and so forth. Although the author is British, the material applies directly to the United States. Appended are a chart of conversion factors from metric units to imperial units and a list of common and botanical names of grasses and plants commonly used in soil conservation. Indexed.

5-238 Hungerford, Harold. *Ecology: The Circle of Life.* Chicago: Childrens Press, 1971. 92p.

This is a simple but complete explanation of basic principles of ecology. Scientific terms are carefully defined in a text that induces concern without moralizing. Illustrated, with a glossary. Indexed. Grades 5-8.

5-239 Hutchins, Carleen Maley. *Who Will Drown the Sound?* New York: Coward, McCann, 1972. 44p.

Similar in style and scope to Thomas and Gretchen Perera's *Louder and Louder* (5-365), this book is even simpler in language and content. It discusses the nature of sound and draws the reader's attention to sounds around him. More emphasis here on industrial and technological design to absorb and reflect sound. Illustrated throughout by Arthur Schaffert. Grades 2-4.

5-240 Hyde, Margaret O. *For Pollution Fighters Only.* New York: McGraw-Hill, 1971. 157p.

An action-oriented book for young people, this text surveys the types of pollution occurring in the United States and suggests ways young citizens can help fight them. The book includes a state-by-state list of pollution control agencies, a list of organizations which provide information on pollution control and careers in the field, and a bibliography. Illustrated with line drawings. Indexed. Grades 7+.

5-241 Hyde, Margaret O. *This Crowded Planet.* New York: McGraw-Hill (Whittlesey House), 1961. 159p.

Although this book was published more than a decade ago, it remains a valid outline of the problems of overpopulation. The author asks basic questions which remain unanswered today and describes two schools of thought—the scientists who feel population must be controlled and those who feel technology can solve the potential problems of an increasing population. Illustrated with line drawings. Indexed. Grades 5-7.

5-242 Hynes, Hugh Bernard Noel. *The Ecology of Running Waters.* Toronto: University of Toronto Press, 1970. 555p.

A comprehensive and critical review of "the biology of rivers and streams." Illustrated extensively with an index to organisms and a truly exhaustive bibliography. Final chapters are devoted to the ecosystem and the effects of man on water courses. Indexed.

5-243 *Indicators of Environmental Quality; proceedings of a symposium held during the AAAS meeting in Philadelphia, Pennsylvania, Dec. 26–31, 1971.* Edited by William A. Thomas. New York: Plenum, 1972. 275p.

A survey of the quantitative patterns of measuring environmental quality in its physical, chemical, and biological aspects. One author deals with the social aspects of the environment. Illustrated. Indexed.

5-244 *Industrial Pollution Control Handbook.* Edited by Herbert F. Lund. New York: McGraw-Hill, 1971. var. paging.

A handbook, by various authors, on pollution control in industry. Chapters on the history of federal pollution-control legislation, state and local laws, and research programs for air and water pollution control. Individual chapters deal with the problems in the steel, chemical, food, pharmaceutical, paper, aerospace, and electronics industries. There is a lengthy section on pollution control equipment and its operation. Includes glossaries for air pollution control and water pollution control. Indexed.

5-245 International Conference on Environmental Future, 1st, Jyvaskyla, Finland, 1971. *The Environmental Future.* Edited by Nicholas Polunin. London: Macmillan (London), 1972. 660p.

The following topics were among those considered by a distinguished list of internationally-known scholars: the biosphere today, monitoring the atmospheric environment, effects of pesticides, soil preservation, pollution and water supplies, and pollution and marine productivity. There are chapters on "global responsibility" and what organizations and industry should do. Conference participants are listed in the book, as well as the conference committees. Papers and the discussions afterwards are recorded. A number of Americans were involved in this conference, including Stewart Udall, David Brower, Reid Bryson, and Charles F. Wurster, Jr. Indexed.

5-246 Irving, Robert McCardle and Priddle, George B., eds. *Crisis: Readings in Environmental Issues and Strategies.* New York: St. Martin's, 1971. 354p.

A collection of readings compiled by Canadians, though a great many of the essays pertain to the United States. The book focuses on the problems of the environment; water, air, and pesticides are treated, and one section is devoted to the "built" environment. A final part of the book deals with solutions to environmental problems. A list of contributors, with biographical information, concludes the volume. Not indexed.

5-247 Ivany, J. W. George. *Environment: Readings for Teachers.* Reading, MA: Addison-Wesley, 1972. 287p.

Readings on the environmental crisis, pollution, human ecology, human beings and radiation, the environment and social action, and the environment and education. Illustrated. An appendix contains recommendations for further reading.

5-248 Jackson, Barbara Ward and Dubos, René Jules. *Only One Earth: The Care and Maintenance of a Small Planet.* New York: Norton, 1972. 225p.

An unofficial report prepared for the United Nations Conference on the Human Environment, Stockholm, 1972. It was prepared with the aid of a 152-member committee of corresponding consultants from 58 countries. The Albert Schweitzer Chair at Columbia University, the World Bank, and the Ford Foundation funded the report. Useful information can be found in the text. An abbreviated index.

5-249 Jackson, Wes. *Man and the Environment.* 2d ed. Dubuque, IA: Brown, 1973. 322p.

The essays and statements in this collection vary in length from a few paragraphs to fifteen pages. They are organized under the headings human behavior, pollution, resources, population and food, and the future. A section on energy includes an interesting taxonomy of international environmental problems and solutions.

5-250 Jennings, Gary. *The Shrinking Outdoors.* Philadelphia: Lippincott, 1972. 191p.

This is a survey of environmental destruction caused by human wastefulness and ignorance. The author discusses air pollution, water pollution, endangered species, and overpopulation, among other topics. Appended are a brief bibliography and list of organizations. Indexed. Grades 7-9.

5-251 Joffe, Joyce. *Conservation*. Garden City, NY: Natural History, 1969. 188p.

This is a young-adult title prepared specifically for the junior-high level. Opening chapters offer a brief history of conservation and the fundamental principles of ecology. Basic terms are carefully explained. Other chapters deal with man's place in nature, the disappearance of wildlife, the importance of wildlife reserves and parks, the population explosion, and such immediate problems as food production and urban design. This is one volume in the publisher's Nature and Science Library series. Profusely illustrated with both black-and-white and color photographs. Indexed. Grades 7+.

5-252 Johnson, C. T. *The Natural World: Chaos and Conservation*. New York: McGraw-Hill, 1971. 288p.

This is a collection of articles reprinted from periodicals and monographs. Part 1 is a group of readings about the environmental crisis by such noteworthies as Rachel Carson and Barry Commoner. Part 2 contains brief glimpses of nature, with essays on the frog, the mackerel, and the desert by Rachel Carson and William O. Douglas, among others.

5-253 Jones, Holway R. *John Muir and the Sierra Club: The Battle for Yosemite*. San Francisco: Sierra Club, 1965. 207p.

This book traces the early development of Yosemite, "as background for the founding of the Sierra Club two years after the establishment of the national park." It ends with the unsuccessful fight to preserve Hetch Hetchy valley in California. A very attractive book, illustrated with black-and-white photographs, line drawings, and maps. Footnotes appear in a narrow column on the inside edge of each page. Several appendixes, containing the original articles of incorporation of the Sierra Club and a list of original members, the first by-laws of the Club, reprints from newspapers of that time, issues facing the newly-founded Sierra Club, and the legislation proposed and supported by the Club.

5-254 Jones, Kenneth Lester; Shainberg, Louis W.; and Byer, Curtis O. *Environmental Health*. San Francisco: Canfield, 1971. 118p.

The title of this small book is a bit misleading, for the scope of the material is broader than "health" might imply. The three sections—on man and environment, population dynamics, and conservation—overlap somewhat, but the text would be useful as an introduction to ecological problems for the layman or high-school student. Too basic for the college classroom. Each section concludes with a summary outline and questions for discussion. Glossary. Bibliography. Indexed.

5-255 Kaill, W. Michael and Frey, John K. *Environments in Profile: An Aquatic Perspective.* San Francisco: Canfield, 1973. 206p.

An introduction to aquatic ecology with a great deal of information on biological tests as well as information and illustrations to facilitate species identification. An appendix covers the planning of field trips and details testing equipment and supplies. Bibliography. Indexed.

5-256 Katz, Robert. *A Giant in the Earth.* New York: Stein & Day, 1973. 262p.

The author presents a provocative alternative to gloomy predictions of global catastrophe. Focusing on the agricultural potential of the earth and proven technology, he maintains that the means to feed both present and future world populations is at hand if it is recognized and put to use. Well documented, but written for the layman. Indexed.

5-257 Kavaler, Lucy. *Dangerous Air.* New York: Day, 1967. 143p.

The author presents case histories of illness and discomfort caused by severe air pollution. Chronicling disasters that have occurred in the United States and elsewhere, she discusses the nature of air pollution, its causes, and the effects it may have on humans, animals, plants, and the climate. She suggests ways everyone can help fight pollution. Illustrated with line drawings. Indexed. Grades 7+.

5-258 Kay, David A. and Skolnikoff, Eugene B. *World Eco-crisis: International Organizations in Response.* Madison: University of Wisconsin Press, 1972. 324p.

These essays, first published in the 1972 spring issue of *International Organization*, deal with the role of international institutions in coping with the impact of modern technology on the global environment. Pages 122-134 include two useful charts on the distribution of environmental activities among international organizations and the nature of these activities. Indexed.

5-259 Klausner, Samuel Z. *On Man in His Environment*. San Francisco: Jossey-Bass, 1971. 224p.

This study grew out of the author's work as a sociology/ psychology consultant to Resources for the Future, Inc. It provides, in the author's own words, "a sociological look at the environment instead of an environmentalist look at sociology." The first chapter is devoted to recent man/environment studies, and the environmental problems of air pollution, noise pollution, and outdoor recreation are treated in later sections of this book. A final chapter deals with policy for environmental research, education, and management. Indexed, with an extensive bibliography. Includes a separate index of personal names.

5-260 Klopfer, Peter. *Behavioral Aspects of Ecology*. 2d ed. Englewood Cliffs, NJ: Prentice-Hall, 1973. 200p.

In this book, "comparative psychology, ethology, and ecology converge into a tight illuminating focus on animal behavior." The author confronts questions like "why don't predators overeat their prey?" and "how are species kept distinct?" A college-level text on animal ecology—brief but somewhat technical. Charts and graphs. Lengthy bibliography. Subject and author indexes.

5-261 Knight, Clifford B. *Basic Concepts of Ecology*. New York: Macmillan, 1965. 468p.

This basic text, designed for introductory courses at the college level, offers chapters on statistical procedures in ecology, ecological succession, community ecology, climatology, and population ecology. Illustrated. Indexed.

5-262 Kohn, Bernice. *The Organic Living Book*. New York: Viking, 1972. 91p.

The author covers such topics as organic gardening, conservation, recycling, and living closer to nature. A final chapter advises readers how to conserve materials and avoid pollution. A bibliography is appended. Indexed. Grades 6-9.

5-263 Kormondy, Edward J. *Concepts of Ecology*. Englewood Cliffs, NJ: Prentice-Hall, 1969. 209p.

A college text by a biologist that attempts to present "the

significant concepts of modern ecology in a readable and intelligible way." It is written for "post-general" biology students and contains chapters on the nature of ecosystems, energy flow in ecosystems, biogeochemical cycles and ecosystems, ecology of populations, the organization and dynamics of ecological communities, and ecology and human beings. Each chapter concludes with a short list of recommended readings, while the preface deals with the history of the term *ecology*.

5-264 Kormondy, Edward J. *Readings in Ecology.* Englewood Cliffs, NJ: Prentice-Hall, 1965. 219p.

A compilation of readings for a basic college course in ecology. Certain papers, according to the editor, are suitable for college students in beginning biology or general science and for high-school students in advanced biology. The editor is himself a biologist. The book contains sections on early natural history, the physical and chemical environment, the study of populations and communities, and the concept of the ecosystem. Essays are arranged chronologically by date of authorship. Not indexed.

5-265 Kostelanetz, Richard. *Social Speculations: Visions for Our Time.* New York: Morrow, 1971. 306p.

A collection that "emphasizes alternative ways of considering contemporary problems." In addition to a section on environments (not all of which are physical), there are readings on cities, technologies, and history. R. Buckminster Fuller is treated at length in this volume, and, while most of the selections are not directly related to environmental issues, the solutions proposed to contemporary problems involve ecological concern. A bibliography of further reading is included. Not indexed.

5-266 Krebs, Charles J. *Ecology: The Experimental Analysis of Distribution and Abundance.* New York: Harper, 1972. 694p.

A first-level college text that presents ecology as a quantitative discipline. Chapters on the distribution of populations and human ecology. Glossary, lengthy bibliography, and species index. Illustrated. Indexed.

5-267 Krutch, Joseph Wood. *The Great Chain of Life.* Boston: Houghton Mifflin, 1957. 227p.

Though almost twenty years old, this book of popular philosophy is interesting reading for a beginner. The author reflects upon the relationships between human beings and animals and describes the complexities of animal evolution, adaptability, and behavior. A worthwhile secondary source, useful at the high-school or lay level.

5-268 Lauwerys, Joseph Albert. *Man's Impact on Nature.* Garden City, NY: Natural History, 1970. 188p.

This volume was published for the American Museum of Natural History. A young-adult title, the book covers such topics as man as hunter, herdsman, and cultivator; use and misuse of land; the machine age; the poisoned environment; and overpopulation. Well illustrated with color and black-and-white photographs and diagrams. Indexed. Grades 7+.

5-269 *Law and the Environment.* Edited by Malcolm F. Baldwin and James K. Page, Jr. New York: Walker, 1970. 432p.

The proceedings of a two-day conference on law and the environment held by The Conservation Foundation in September 1969. A major section is devoted to "needed developments in the law." Papers presented in the collection were prepared by attorneys. A lengthy bibliography on environmental law is included. Indexed.

5-270 Laycock, George. *Alaska: The Embattled Frontier.* Boston: Houghton Mifflin, 1971. 205p.

This is the first volume of The Audubon Library, a series published in cooperation with the National Audubon Society. The author, a field editor for *Audubon* magazine, describes Alaska's land and people, past conservation battles, resource management, present dangers to Alaskan wilderness and wildlife, and the dangers of oil exploration and transport. Indexed. Illustrated.

5-271 Laycock, George. *America's Endangered Wildlife.* New York: Norton, 1969. 226p.

Written for young adults, and amply illustrated with photographs, this book provides an interesting survey of fifteen different species of wildlife in danger of becoming extinct. It contains a list of organizations concerned with endangered wildlife and a bibliography. The U. S. Fish and Wildlife Service's list of rare and endangered species is reprinted. Indexed. Grades 7+.

5-272 Laycock, George. *The Diligent Destroyers*. Garden City, NY: Doubleday, 1970. 225p.

The author, a naturalist, writes a documented account of the destruction wrought by "the dam-builders, the strip miners, the highway engineers, and their self-serving allies." Coming down hard on engineers, and the U.S. Army Corps of Engineers in particular, the author presents case histories of their rush to "remodel America" and (according to Laycock) destroy it. A final chapter offers "cures." Illustrated. Selected bibliography. Indexed.

5-273 Leaf, Munro. *Who Cares? I Do*. Philadelphia: Lippincott, 1971. 40p.

Through a simple text, cartoon characters, and photographs, the author attempts to raise the consciousness of young children concerning littering and despoiling the environment. The villains are "droppers," "spoilers," and "wreckers." Grades K-3.

5-274 Leinwand, Gerald and Popkin, Gerald, eds. *Air and Water Pollution*. New York: Washington Square, 1969. 160p.

One of the series Problems of American Society, this book is designed as an introductory text for urban schools. It was partly funded under Title I of ESEA. The book includes articles by Ben Bagdikian, Rachel Carson, and Robert Kennedy. Each essay is followed by several questions for further discussion. Illustrated. Indexed.

5-275 Leithe, Wolfgang. *The Analysis of Air Pollutants*. Ann Arbor, MI: Ann Arbor Science, 1971. 304p.

A text, first published in German in 1968, which presents the methodology and techniques for analysis of air pollutants, with new addenda for the English edition. Bibliography. Indexed.

5-276 Leopold, Aldo A. *Sand County Almanac*. New York: Oxford, 1966. 269p.

A philosophical statement by a conservationist, who defines nature as "the community to which we belong . . . not a community which belongs to us." Chapters on the "conservation esthetic," wildlife in American culture, wilderness, and the "land ethic" highlight this classic. Excellent nature writing. Not indexed.

5-277 Lewis, Howard R. *With Every Breath You Take; the poisons of air pollution, how they are injuring our health, and what we must do about them.* New York: Crown, 1965. 322p.

The author, a public health consultant, hopes to alert the layman to the dangers of air pollution. The book focuses on the effects of air pollution upon health, documenting tragic "incidents," detailing long-term effects, and discussing possibilities of control. Appendixes include a brief history of the scholarly study of air, sampling techniques, and suggestions for further reading. Illustrated. Indexed.

5-278 Lewis, Trevor and Taylor, L. R. *Introduction to Experimental Ecology.* New York: Academic, 1967. 401p.

Two British ecologists present the principles of ecology in a simple quantitative manner. There are forty-five exercises in ecology and a chapter on "keys to common land invertebrates" to assist the student in identifying specimens. Illustrated. Indexed.

5-279 Leydet, François. *The Last Redwoods, and the Parkland of Redwood Creek.* San Francisco: Sierra Club, 1969. 160p.

Based on *The Last Redwoods*, published by the Sierra Club in 1963, this volume has been updated to reflect the 1968 law creating the Redwood National Park. An ecological history of the great redwoods and an analysis of their collective "life expectancy," this book is as noteworthy for its exquisite color plates as for its text. Includes a species list of flora and fauna illustrated in the volume. Illustrated.

5-280 Line, Les. *What We Save Now: An Audubon Primer of Defense.* Boston: Houghton Mifflin, 1973. 438p.

This book consists of essays which appeared in recent issues of *Audubon Magazine* and were brought together by the editor of that journal. The essays are arranged in three sections and feature some of the most prominent conservation writers active today. Illustrated with photographs. Biographical notes on contributors are contained in a separate section. Indexed.

5-281 Lineberry, William P. *Priorities for Survival.* New York: Wilson, 1973. Reference Shelf, Volume 44, no. 6. 223p.

A collection of readings, primarily from popular periodicals, which present a cross-section of views on the environmental crisis, its relation to technology and to the economy, and some possible solutions. Substantial bibliography.

5-282 Linton, Ron M. *Terracide: America's Destruction of Her Living Environment.* Boston: Little, Brown, 1970. 376p.

A popularly written but substantial book about the state of the environment—how it got that way and what to do about it. The first section deals with the physical and psychological effects of all kinds of pollution on humans. The second section recounts commercial and industrial sins and details how technology has affected the environment. The author's concluding section contains a description of the government's ineffectiveness in combatting pollution, along with suggestions for solving environmental problems. Bibliography. Indexed.

5-283 Livingston, John A. *One Cosmic Instant: Man's Fleeting Supremacy.* Boston: Houghton Mifflin, 1973. 243p.

This popular monograph is concerned with the "ill-balanced relationship" between man and nature. The author selectively chronicles both the evolution of the world and the evolution of human thinking about the world. Livingston hopes for a new environmental ethic which will no longer egotistically separate the human race from the rest of nature. Bibliography. Indexed.

5-284 Loftas, Tony. *The Last Resource: Man's Exploitation of the Oceans.* New rev. ed. Chicago: Regnery, 1970. 276p.

This book carefully describes the value of the ocean as a resource for food, water, minerals, and energy. It deals briefly with pollution, but concentrates on problems of national and international control of the ocean, with a separate chapter on "wars and wayfaring." Glossary. Illustrated. Indexed.

5-285 Longgood, William Frank. *The Darkening Land.* New York: Simon & Schuster, 1972. 572p.

According to the introduction, "this book attempts to explain the world's intricate interrelationships and how pollutants interfere with the life-support systems. Its scope embraces not only environmental damage and the peril it poses to life but also political, social, and economic influences." An appendix contains a listing and descriptions of the common environmental poisons. Extensive bibliography. Indexed.

5-286 Loraine, John Alexander. *The Death of Tomorrow*. Philadelphia: Lippincott, 1972. 376p.

A book by an English doctor which deals with the environment from a historical perspective and suggests that "unless we act now, the prognosis for tomorrow will be bleak indeed." Published originally by William Heineman Medical Books, this volume deals with both the United States and the United Kingdom, but its main focus is on acute environmental problems abroad.

5-287 Loth, David G. and Ernst, Morris L. *The Taming of Technology*. New York: Simon & Schuster, 1972. 256p.

This book "outlines the legal restrictions that can be used to curb the environmentally destructive aspects of technological growth." The authors consider the legal aspects of energy, water, air, ocean, and space, and branch out into issues of law in medicine and in the computer field. A table of cases and list of sources are included at the end of the volume. Indexed.

5-288 Love, Glen A. and Love, Rhoda M., eds. *Ecological Crisis: Readings for Survival*. New York: Harcourt, 1970. 342p.

A general collection of reprinted essays on the environment, covering both scientific and political issues. Authors include Paul Ehrlich, Gaylord Nelson, Barry Commoner, Rachel Carson, William O. Douglas, and Aldous Huxley. Bibliography included.

5-289 Lynch, Patricia and Chandler, Robert. *National Environment Test*. New York: Pocket, 1971. 127p.

The National Environment Test was adapted from a program presented by CBS News. There are six sections to the test, and the results of the CBS News Poll are included. There is a brief final chapter entitled "Where Are We Going." Illustrated. Not indexed.

5-290 McCamy, James L. *Quality of the Environment*. New York: Free Press, 1972. 276p.

Written by a political scientist for the layman, this book discusses the state of the environment and how institutions can be used to improve it. The book is a rewritten version of an interdisciplinary seminar held at the University of Wisconsin in 1967. Part 1 describes the condition of the total environment of the

United States, and focuses on the environment of two counties in Wisconsin, one urban and one rural. Part 2 analyzes institutions, such as government, business, and the professions, and suggests ways to use them effectively to achieve change. Indexed.

5-291 McClellan, Grant S. *Protecting Our Environment.* New York: Wilson, 1970. Reference Shelf, volume 42, no. 1. 218p.

A collection of essays in the Reference Shelf series that deals with protecting our environment. A final section, entitled "What Can Be Done?" considers Sweden's antipollution program and several approaches to combatting pollution in the United States. Bibliography.

5-292 McCluney, William Ross. *The Environmental Destruction of South Florida: A Handbook for Citizens.* Coral Gables, FL: University of Miami Press, 1969. 134p.

A compilation of readings on the environment of South Florida. Some of the material is reprinted, but a substantial portion has been written expressly for this book. Articles on air and water pollution, architecture, population, and other topics. A selected bibliography is included. Not indexed.

5-293 McCoy, Joseph. *Saving Our Wildlife.* New York: Crowell-Collier, 1970. 232p.

The history of conservation, especially of wildlife, in America. The author begins in colonial times and traces the economic and social factors which have led to the extinction or near-extinction of many species. He also surveys the current situation and offers suggestions for future wildlife management. Includes glossary, bibliography, and a brief list of environmental organizations. Illustrated with black-and-white photographs. Indexed. Grades 6-8.

5-294 McCoy, Joseph. *Shadows Over the Land.* New York: Seabury, 1970. 152p.

In a survey of the present environmental crisis, the author discusses air and water pollution, endangered wildlife and wilderness, the dangers of pesticides, and urban blight. The author suggests ways in which young people can help remedy the situation. Appended are a glossary, list of suggested readings, recommended films, and a list of names and addresses of conservation organizations. Indexed. Grades 3-6.

5-295 McHale, John. *The Ecological Context.* New York: Braziller, 1970. 188p.

The book is a revision of the author's document #6 (1967), one of a series of reports issued by the World Resources Inventory at Southern Illinois University. Illustrated with numerous charts, it attempts to provide an insight into the efficient and ecologically sound use of natural resources and energy production through technological development. Not indexed, but there is a lengthy bibliography. There is a "Chart Reference List" of citations as well.

5-296 McHale, John. *World Facts and Trends.* 2d ed. New York: Macmillan, 1972. 95p.

According to the author, this second edition has been "reorganized so that it may be used as an introductory text focusing on the interrelationships of key trends at the world level." The book is about half text and half charts and graphs, with interpreted statistical data on human beings and the biosphere, environmental systems, human systems, and comparative indicators. Data include, for example: world consumption of calories and protein by country, water consumption patterns in the United States, and composition of the biosphere by element. The book contains 130 line drawings and graphs with chart and text references appearing at the end of the book. Briefly indexed.

5-297 McKain, David W. *The Whole Earth: Essays in Appreciation, Anger, and Hope.* New York: St. Martin's, 1972. 276p.

The editor has chosen essays which are "uncompromisingly personal, even intimate." The first eleven selections are word pictures of moments in nature by men such as D. H. Lawrence, Ralph Waldo Emerson, and George Orwell. The next fifteen pieces are angry cries against the destruction of nature, and the final essays offer some hope for reconciliation between humankind and nature.

5-298 McKenzie, Roderick Duncan. *Roderick D. McKenzie on Human Ecology: Selected Writings.* Edited by Amos H. Hawley. Chicago: University of Chicago Press, 1968. 308p.

The selected writings of a distinguished sociologist on human ecology. McKenzie first conceived of human ecology as "a study of the spatial and temporal relations of human beings as affected by the selective, distributive and accommodative forces of the environment," but later altered this definition to place less emphasis on spatial patterns. Useful as a document in the history of sociological thought, although not of current ecological interest. Not indexed.

5-299 Mackintosh, Douglas R. *The Economics of Airborne Emissions: The Case for an Air Rights Market.* New York: Praeger, 1973. 121p.

A proposal to establish an air rights market as the solution to air pollution. According to the author, "in order to institute an air rights market, an agency (public, nonprofit, or proprietary) would have to perform two primary functions: (1) provide a market for the exchange of air rights, and (2) police individual emissions. . . ." The agency would sell an initial offering of air rights to begin the program. Not indexed, but a lengthy bibliography accompanies this study. The author, at the time the study was published, was a professor of business administration.

5-300 McNulty, Faith. *Must They Die? The Strange Case of the Prairie Dog and the Black-Footed Ferret.* Garden City, NY: Doubleday, 1971. 86p.

This is a story of "prairie dogs, ferrets, and bureaucrats." The author documents "governmental schizophrenia"—the conflict of purposes within the Bureau of Sport Fisheries and Wildlife, which seeks to both conserve and destroy wildlife. This case history presents a perfect example of the problems of protecting wild animals that are not economically valuable. Illustrated with black-and-white photographs.

5-301 Maddox, John. *Doomsday Syndrome.* New York: McGraw-Hill, 1972. 248p.

According to the author, this book is "not a scholarly work but a complaint." Maddox, editor of *Nature* magazine, objects to what he feels are exaggerations of the environmental crisis by many authors of recent articles and books. One by one he takes them to task for misusing statistics, overdramatizing pollution, and ignoring the possibilities of prosperity without catastrophe.

5-302 *Man and the Ecosphere: Readings from "Scientific American."* With commentaries by Paul R. Ehrlich, John P. Holdren, and Richard W. Holm. San Francisco: Freeman, 1971. 301p.

Essays reprinted from *Scientific American* and grouped around the following topics: the ecosphere and preindustrial man, limits rarely perceived, the dimensions of intervention, and on management and buying time. There are biographical notes about contributors, with additional bibliographies. Well illustrated. Indexed.

5-303 *Managing the Environment: International Cooperation for Pollution Control.* Edited by Allen V. Kneese and others. New York: Praeger, 1971. 356p.

The proceedings of the conference, Goals and Strategy for Environmental Quality Improvement in the 1970s, organized by The Atlantic Council and Battelle Memorial Institute. Financial assistance was provided by IBM and the Allegheny Foundation. The conference participants consisted primarily of industrialists from North America, Europe, and Japan. Included are presentations by Russell E. Train on "Goals and Strategy for Environmental Quality Improvement in the United States" and a paper by the Battelle Memorial Institute on "Environmental Quality: The Technological State of the Art." An appendix, with various supplementary papers, is included, as is a list of conference participants.

5-304 Margalef, Ramon. *Perspectives in Ecological Theory.* Chicago: University of Chicago Press, 1968. 111p.

This book consists of four lectures delivered in the Department of Zoology at the University of Chicago in May 1966. Essays are on the ecosystem as a cybernetic system, ecological succession and exploitation by human beings, the study of pelagic ecosystems, and evolution in the frame of ecosystem organization. A short bibliography concludes the book. Indexed.

5-305 Marine, Gene. *America the Raped: The Engineering Mentality and the Devastation of a Continent.* New York: Simon & Schuster, 1969. 331p.

The author, a journalist, documents in very readable prose endangered ecosystems and natural beauty spots in America. His anger is directed particularly at engineers who "commit rape with bulldozers." Name and subject indexes.

5-306 Marine, Gene and Van Allen, Judith. *Food Pollution: The Violation of Our Inner Ecology*. New York: Holt, 1972. 385p.

Written in a popular style, this book supplies basic information about the possible health hazards of food additives, food colors, DDT, and other pesticides. The book contains extensive notes and a bibliography. Indexed.

5-307 Marx, Wesley. *The Frail Ocean*. New York: Coward, McCann, 1967. 274p.

An in-depth popular account of the ocean's potential and its precarious state. The author chronicles the biological damage already done, environmental disasters and their aftermath, and the search for means to govern the ocean wisely. Suggested references are included at the end of the volume. Among several appendixes are: comparative annual production, live weight of animals in pounds per acre, the presidential ocean program for 1968, and seaward claims by nations. Illustrated.

5-308 Marx, Wesley. *Man and His Environment: Waste*. New York: Harper, 1971. 179p.

This book deals with the problems created by using the planet as a dump. The economics and technology of waste management, the health dangers of increasing waste loads, a recommendation for "closed system" recycling, and a discussion of adaptation to pollution versus control of pollution are included in the text. Suitable for classroom or community use. Each chapter concludes with "suggested readings." Illustrated. Indexed.

5-309 Marx, Wesley. *Oilspill*. San Francisco: Sierra Club, 1972. 139p.

This book goes beyond marine ecology to cover national and international economics and politics which affect marine life. The author feels that "The United States is on a hydrocarbon high without knowing oil's true price. . . ." Highly critical of the oil industry, the author outlines the dangers of oil spills to the ocean and to people, the ineffectiveness of present regulation procedures, legal ramifications of spills, and a perspective on needed changes. Not indexed.

5-310 Marzani, Carl. *The Wounded Earth*. Reading, MA: Young Scott, 1972. 232p.

This is a survey of the earth's environment and environmental problems. The first section of the book discusses basic ecological concepts such as ecosystems. Part 2 deals with pollution of all kinds and the third section offers some solutions to the problems. Few illustrations. Bibliography included. Indexed. Grades 8+.

5-311 Matthews, William H. *Man's Impact on Terrestrial and Oceanic Ecosystems.* Cambridge: MIT Press, 1971. 540p.

This book and its companion volume, *Man's Impact on the Climate* (5-312), were prepared from the background material and working papers of the 1970 interdisciplinary Study of Critical Environmental Problems (SCEP) sponsored by MIT. The book expands upon the published report of SCEP, *Man's Impact on the Global Environment* (1-6). The thirty-three papers collected here include overviews of ecological problems resulting from technology and overpopulation, the direct and indirect effects of pollution on terrestrial and marine ecosystems, measurement and monitoring systems, mathematical models, and relevant social and political issues. Each article includes a list of references. Name and subject index.

5-312 Matthews, William H.; Kellog, William W.; and Robinson, G.D., eds. *Man's Impact on the Climate.* Cambridge: MIT Press, 1971. 594p.

This book contains forty-eight papers and is divided into eleven parts. Each section contains an introduction. Part 1 carries the SCEP (Study of Critical Environmental Problems) Work Group reports on the climate and atmospheric monitoring. Part 2 provides a semitechnical view of the factors which can change or determine the climate. Parts 5 through 9 provide discussions of specific pollutants that may affect the environment. Monitoring techniques are discussed in part 10 and part 11 addresses political and social issues related to the climate and the environment. This volume and *Man's Impact on Terrestrial and Oceanic Ecosystems* (5-311) contain background materials to *Man's Impact on the Global Environment* (1-6), the SCEP study. In addition, both contain papers written during the study, as well as a few essays which were previously published. Most of the material is technical, but it is nonetheless essential to an environmental collection. Illustrated. Indexed.

5-313 Meadows, Donella H.; Meadows, Dennis L.; Ranclus, Jørgen; and Behrens, William W., III, eds. *The Limits of Growth: A Report for the Club of Rome's Project on the Predicament of Mankind.* New York: Universe Books, 1972. 205p.

A group of about thirty met in April 1968 in Rome where this project was launched. The present book results from the work of a Massachusetts Institute of Technology project team, directed by Professor Dennis Meadows, and Potomac Associates, a DC-based research organization. Chapters on the nature of exponential growth, the limits of this growth, growth in the world system, technology, and the state of global equilibrium. Commentary by the Club of Rome Executive Committee is included, as is a short bibliography of related studies. Not indexed. Illustrated. (See 5-314.)

5-314 Mesarovic, Mihajlo and Pestel, Eduard. *Mankind at the Turning Point: The Second Report to the Club of Rome.* New York: Dutton, 1974. 210p.

An examination of developing world crises in natural resources, food supplies, and population. Amply illustrated. Indexed. (See 5-313.)

5-315 Meyer, Alfred. *Encountering the Environment.* New York: Van Nostrand Reinhold, 1971. 212p.

A group of articles reprinted from *Natural History*, the journal of the American Museum of Natural History. All the essays are concerned with some aspect of the environment. Not indexed.

5-316 Migel, J. Michael. *The Stream Conservation Handbook.* New York: Crown, 1974. 242p.

Written for fishermen concerned about declining fishing grounds, this book is nonetheless relevant for all interested laymen. Chapters on stream ecology, the stream killers, stream-side surveillance, and stream improvement; also, two chapters on group action and legal action to protect and save streams. Each chapter is separately authored. Includes a selected list of relevant organizations and a selected bibliography. Illustrated. Indexed.

5-317 Miller, G. Tyler. *Replenish the Earth: A Primer in Human Ecology.* Belmont, CA: Wadsworth, 1972. 190p.

Intended for use by high-school or college students, this book explains basic principles and ecological concepts. Sections on population dynamics, the first and second laws of thermodynamics, cybernetics, world hunger, pollution and technology, and overpopulation. Concludes with recommendations for action, both for society and for the individual. Serious but readable presentation of technical data. Charts and graphs.

5-318 Milne, Lorus and Milne, Margery. *Arena of Life: The Dynamics of Ecology.* Garden City, NY: Doubleday, 1972. 350p.

A profusely illustrated textbook on ecology, useful from the high-school level up. Explanations of basic natural processes such as the carbon cycle and parasitism lead to more specific treatment of the human environment and its present decay. Separate chapters on the ecology of various environments— marine, fresh water, soil, forest, grassland, desert, and polar. Appendix outlines the binomial classification system. Glossary. Bibliography by chapter. Indexed.

5-319 Milne, Lorus and Milne, Margery. *The Nature of Life: Earth, Plants, Animals, Man and Their Effect on Each Other.* New York: Crown, 1970. 320p.

The authors present an overview of the natural history of the world, including evolution and patterns in nature. After opening chapters on the nature of the earth and how it changes, the book treats the geographical areas of the world and their species. Lavishly illustrated with color and black-and-white photographs. Detailed index to animals and plants included.

5-320 Milne, Lorus and Milne, Margery. *Patterns of Survival.* Englewood Cliffs, NJ: Prentice-Hall, 1967. 339p.

A book about the techniques of survival found in plants and animals, which explains why some species have survived and others not. Chapters on such subjects as how to avoid being eaten, various types of armor and other protective coverings, and how to survive a drought. Indexed. Illustrated.

5-321 Mines, Samuel. *The Last Days of Mankind.* New York: Simon & Schuster, 1971. 319p.

A book on the seriousness of our environmental problems, with sections on water and wetlands, woodlands, wildlife, civilization, and land in the public domain. Indexed.

5-322 Mitchell, Ralph. *Water Pollution Microbiology.* New York: Wiley–Interscience, 1972. 416p.

A book that attempts to assess the role of microorganisms in water pollution and their control. Essays, contributed by different authors, on microbial changes induced by inorganic pollutants, microbial changes induced by organic pollutants, intestinal pathogens as pollutants, pollution and community ecology, microbial parameters of pollution, and microbiological approaches to pollution control. Indexed.

5-323 Monsen, R. Joseph. *Business and the Changing Environment*. New York: McGraw-Hill, 1973. 290p.

The word *environment* in the title of this book is used to mean *society* or *external factors*. Thus the author is concerned with pressures on business from minority groups and consumers as well as pressure from environmental regulations. In addition, the book provides historical perspectives on the development of business in the United States, the capitalistic system, and power groups. The author concludes with a section on values—past, present, and future—as they are part of the business ethic. Each chapter is followed by several discussion questions. Indexed.

5-324 Montague, Katherine and Montague, Peter. *Mercury*. San Francisco: Sierra Club, 1971. 158p.

The authors present a terrifying account of mercury poisoning—its long-term effects, how mercury gets into our food, why its dangers were not known and controlled sooner, and case histories of people affected by it. The authors document governmental inaction in the face of known dangers. A lengthy postscript of other potentially dangerous substances now common in our environment is added. Appendixes include a list of industrial uses of mercury, a state-by-state pollution survey, and workplace standards. Selected bibliography.

5-325 Morgan, Arthur E. *Dams and Other Disasters*. Boston: Sargent, 1971. 422p.

A history of environmental disasters connected with the U.S. Army Corps of Engineers' civil works projects over the past century. The book is introduced by Paul H. Douglas, former senator from Illinois. Indexed.

5-326 Murdoch, William W. *Environment: Resources, Pollution, and Society*. Stamford, CT: Sinauer, 1971. 440p.

A text on ecology written by twenty-one authors with sections on population and resources, environmental degradation, and environment and society. Each chapter contains recommendations for further reading. Indexed.

5-327 Murphy, Earl Finbar. *Governing Nature.* Chicago: Quadrangle, 1967. 333p.

This book is an examination of the origins of environmental control and a history of the principles of its current application. The author, an attorney, draws upon legal, historical, and contemporary sources to provide an explanation and a rationale for the regulation of resources in our environment. The book contains extensive chapter notes and a bibliography. Indexed.

5-328 Murphy, Earl Finbar. *Man and His Environment: Law.* New York: Harper, 1971. 168p.

This book by a law professor at Ohio State University deals with the philosophy and history of law as it has been applied to the environment. The author discusses past and current laws, but he is more interested in legal concepts than in specific laws. The book shows how the law's "essentially neutral quality can further either exploitation or preservation of our natural resources." List of references with each chapter. Indexed.

5-329 National Academy of Sciences, Washington, D.C. Office of the Foreign Secretary. *Rapid Population Growth: Consequences and Policy Implications.* Baltimore: Johns Hopkins University Press, 1971. 696p.

This comprehensive analysis of population growth is divided into two parts. The first section is an overview which "relates technical findings to public policy on population growth and makes specific recommendations for future population control." The second section includes seventeen technical papers on such topics as health care, education, family size, and resource adequacy. Charts and graphs. Indexed.

5-330 National Conference on Control of Hazardous Materials Spills, University of Houston, 1972. *Control of Hazardous Materials Spills; Proceedings.* Washington, DC: Graphics Management, 1972. 223p.

Sponsored by the Environmental Protection Agency, this conference appears to have been a discussion among governmental officials and representatives of major industries. Papers included deal with the prevention of hazardous materials spills in heavy industry, planning for response to spills, containment of spilled materials, detection and identification of spills, effects of spills on the environment, and similar topics. Illustrated.

5-331 National Conference on Managing Irrigated Agriculture to Improve Water Quality, Colorado State University, 1972. *Managing Irrigated Agriculture to Improve Water Quality; Proceedings.* Washington, DC: Graphics Management, 1972. 306p.

This volume includes twenty-eight technical papers presented at a conference sponsored by the Environmental Protection Agency and Colorado State University in May 1972. Each paper is preceded by its own abstract. Illustrated. Not indexed.

5-332 National Conference on UNESCO: 13th. San Francisco, 1969. *No Deposit, No Return. Man and His Environment: A View Toward Survival.* Edited by Huey D. Johnson. Reading, MA: Addison-Wesley, 1970. 351p.

Compiled for the layman, this anthology is, as the introduction states, a "primer for environmental awareness," and an excellent one. A distinguished list of scholars, writers, and citizens contributed. It is not indexed, but carries rather extensive biographical notes about all contributors.

5-333 National Research Council. Committee on Biologic Effects of Atmospheric Pollutants. *Asbestos: The Need for and Feasibility of Air Pollution Controls.* By the Committee on Biological Effects of Atmospheric Pollutants, Division of Medical Sciences, National Research Council. Washington, DC: National Academy of Sciences, 1971. 40p.

This study sets forth the information available at the time of preparation on asbestos as an air pollutant, and its effects upon man. The study was prepared by the Air Pollution Control Office of the Environmental Protection Agency. A bibliography of references cited in the study concludes the volume. Not indexed.

5-334 National Research Council. Committee on Geological Sciences. *The Earth and Human Affairs.* San Francisco: Canfield, 1972. 139p.

Written by a committee of geologists under the auspices of the National Academy of Sciences and the National Research Council, this book is a general outline of the problems and processes of the environmental crisis from the point on view of one scientific specialization. Chapters on energy resources, radioactive wastes, salt, copper, geologic hazards, earth history, mines and quarries, groundwater, wastes and pollution, and many other topics. Brief list of suggested readings. Indexed. Illustrated.

5-335 National Research Council. Committee on Resources and Man. *Resources and Man: A Study and Recommendations.* By the Committee on Resources and Man of the Division of Earth Sciences, National Academy of Sciences—National Research Council. San Francisco: Freeman, 1969. 259p.

The result of a two-year study, this book opens with the committee's twenty-six recommendations for action. The suggestions deal with the areas of concern explored more fully in the remainder of the book, namely, population, food, mineral resources, and energy resources. Each chapter concludes with a list of references. Indexed.

5-336 National Symposium on Thermal Pollution, Portland, OR, 1968. *Biological Aspects of Thermal Pollution: Proceedings.* Edited by Peter A. Krenkel and Frank L. Parker. Nashville, TN: Vanderbilt University Press, 1969. 407p.

Proceedings of a symposium sponsored by the Federal Water Pollution Control Administration and Vanderbilt University, the papers cover such topics as the engineering aspects, sources, and magnitude of thermal pollution; research needs for thermal pollution control; and the effects of thermal pollution on certain species of plant and animal life. Most papers include a list of references and a transcript of the discussion which followed presentation. A list of participants in the discussion is included at the end of the volume. Illustrated. Indexed.

5-337 *Natural Resources: Quality and Quantity.* Edited by S. V. Ciriacy-Wantrup and James J. Parsons. Berkeley: University of California Press, 1967. 217p.

Papers presented over a period of five years, all dealing with natural resources and one or more environmental problems associated with each resource. Essays by Lewis Mumford, Albert Lepawsky, and others. Illustrated. Indexed.

5-338 Naumov, Nikolai Pavlovich. *The Ecology of Animals.* Edited by Norman D. Levine. Translated by Frederick K. Plous. Urbana: University of Illinois Press, 1972. 650p.

Originally published in the USSR in 1963, this is a standard text on animal ecology in the Soviet Union. The editor points out distinctions and discrepancies between basic American and Russian concepts. The text is divided into three sections, on the ecology of individuals, populations, and associations. Long list of references. Should be used as supplementary material or for comparative study. Indexed.

5-339 Navarra, John Gabriel. *Our Noisy World.* Garden City, NY: Doubleday, 1969. 203p.

A simple text on noise pollution, this book was written for the layman but could be used from the junior-high level on up. Profusely illustrated, the book describes basic principles of sound and energy, alerts the reader to the dimensions of the noise pollution problem, and offers some suggestions for noise control. Indexed. Grades 7+.

5-340 Nelson-Smith, Anthony. *Oil Pollution and Marine Ecology.* New York: Plenum, 1973. 260p.

First published in the United Kingdom, this book deals with oil pollution and its effect on the marine environment. Chapters on coping with spilt oil, sources of oil pollution, and the effects of oil on marine organisms. Long bibliography. Indexed.

5-341 Neuhaus, Richard J. *In Defense of People: Ecology and the Seduction of Radicalism.* New York: Macmillan, 1971. 315p.

This polemic argues against hysteria over the environmental crisis, suggesting that exaggerated fears have short-circuited political processes and diverted attention from vital issues such as social change, feeding the world's hungry, and in the past, the Vietnam War. The author warns against blaming the poor people of the earth for any imminent environmental crisis since, he feels, world poverty is created by the rich people. The author is pastor of the Lutheran Church of St. John the Evangelist in Brooklyn, New York, and a founder of National Clergy and Laymen Concerned About Vietnam. Not indexed.

5-342 Ng, Larry K. Y. and Mudd, Stuart, eds. *The Population Crisis: Implications and Plans for Action.* Bloomington: Indiana University Press, 1970. 276p.

This is a "condensed and reorganized" version of *The Population Crisis and the Use of World Resources,* edited by Stuart Mudd. The original text came out of a collaboration between The World Academy of Art and Science and a group of undergraduate students at Stanford University. It is a collection of essays on the population crisis, covering a wide range of specific topics such as problems of fertility control, population control in China, man and hunger, use and abuse of land, genetic inheritance, and several action programs. Appended are a list of voluntary organizations and research programs, notes, bibliography, and a list of contributors. Indexed.

5-343 Nicholson, Max. *The Big Change: After the Environmental Revolution.* New York: McGraw-Hill, 1973. 288p.

Sequel to *The Environmental Revolution,* this is a where-do-we-go-from-here book on the problems of human social evolution. The author emphasizes the need to shape "the big change," to develop a new habitat for man since the old environment is dissolving before our eyes. Concerned with the internal, psychological environment as well as external forces, he discusses ecological humanism, the population explosion, the prospect of doomsday, and the quality of life. Indexed. (See 5-344.)

5-344 Nicholson, Max. *The Environmental Revolution: A Guide for the New Masters of the World.* New York: McGraw-Hill, 1970. 366p.

A book by the Director-General of the Nature Conservancy. It explores the worldwide problems of humanity's relation to the environment, with attention to the United States as well as Great Britain. Amply illustrated, it carries appendixes on the vegetation cover of the earth, a chart showing human impact on the countryside, and flow charts of conservation processes. Indexed, with an extensive section on sources. (See 5-343.)

5-345 Nybakken, James Willard. *Readings in Marine Ecology.* New York: Harper, 1971. 544p.

Compiled for the advanced undergraduate or the graduate student in marine ecology, this is a collection of articles from various scholarly scientific journals. The papers are organized under the headings benthic ecology, plankton, deep sea ecology, intertidal ecology, tropical ecology, and concepts in marine ecology. The articles are reprinted in full with accompanying illustrations. Each carries a list of literature cited or a bibliography. Not indexed.

5-346 *Ocean Resources and Public Policy.* Edited by Thomas Saunders English. Seattle: University of Washington Press, 1973. 184p.

Based on an interdisciplinary seminar sponsored by the University of Washington Graduate School of Public Affairs, this book includes eleven essays which discuss the natural and physical properties of the ocean, the present state of marine industries, desalination, energy, and so forth, and consider the legal and governmental aspects of ocean resource policy. Indexed.

5-347 Odum, Eugene P. *Ecology.* New York: Holt, 1963. 152p.

This book, written as a college text, presents a series of "graphic models" that illustrate the principles of ecology from the standpoint of the biologist. Chapters on the scope of ecology, the concept of the ecosystem, energy flow, biogeochemical cycles, "limiting factors," ecological regulation, and ecosystems of the world. Each chapter is accompanied by a list of suggested readings. The index provides the locations in the text where terms and concepts are most fully defined or explained.

5-348 Odum, Eugene P. *Fundamentals of Ecology.* 3d ed. Philadelphia: Saunders, 1971. 574p.

A college text, by a zoologist, that in this edition has been expanded into three books: Book 1 is an overview of how ecology relates to human affairs; book 2 is designed for the undergraduate college course in ecology with recommended laboratory and field work; and book 3 is a comprehensive reference work on principles, environments, and ecological technology. Contains an extensive bibliography.

5-349 Odum, Howard T. *Environment, Power, and Society.* New York: Wiley–Interscience, 1970. 331p.

This book demonstrates the application of "basic laws of energy and matter to the complex system of nature and man." The author uses "energy diagraming" to consider the problems of power, pollution, food, and war. He covers such topics as power in ecological systems, a power basis for humanity, power and economics, power and politics, and energy as a basis for religion. Although the author states that the book is for the general reader, it is, by nature of its subject matter, sophisticated reading. Energy module formulas are appended. Charts and graphs. Indexed.

5-350 Ogden, Samuel K. *America the Vanishing: Rural Life & the Price of Progress*. Brattleboro, VT: Greene, 1969. 242p.

An anthology which the editor calls "both a nostalgic memorial to what was, and a disapproving commentary on what is." The first three chapters include essays on the beautiful in America by such well-known people as John James Audubon, Mark Twain, Hamlin Garland, and Henry David Thoreau, plus many lesser-known commentators on the pleasures of nature. Chapters 4 and 5 contain articles about environmental disasters, the "price of progress," and what the future holds. Illustrated.

5-351 Oklahoma University. Science and Public Policy Program. Technology Assessment Group. *Energy Under the Oceans*. Norman: University of Oklahoma Press, 1973. 378p.

A survey of the technology and procedures of undersea oil and gas exploration. The survey was funded by the National Science Foundation at the University of Oklahoma. Policy recommendations are offered in a number of areas. Illustrated. Indexed.

5-352 Olsen, Jack. *Slaughter the Animals, Poison the Earth*. New York: Simon & Schuster, 1970. 287p.

Written in anecdotal prose, this is a story of sheepmen poisoning coyotes. The author is decidedly on the side of the coyote and against sheep, sheepmen, and the U.S. Fish and Wildlife Service. This book presents the conservationist's point of view on the use of lethal poisons, the people who use them against predatory animals, and the true nature of those animals. Grades 7+.

5-353 Oltmans, Willem L. *On Growth*. New York: Capricorn, 1974. 493p.

A sequel to the Club of Rome's *The Limits to Growth* (see 5-313) that features interviews with approximately seventy individuals discussing the report. A broad range of persons from a variety of backgrounds are interviewed. Indexed.

The author has published a supplement, *On Growth II*, which appeared in 1975.

5-354 Opie, John. *Americans and Environment: The Controversy Over Ecology.* Lexington, MA: Heath, 1971. 203p.

A collection of readings about the environment. The first section is devoted to earlier American opinion about nature beginning with Emerson and ending with Aldo Leopold. Subsequent material appears under "The Contemporary Controversy Begins," "Contemporary Environmental Problems," and "Alternative Solutions." Not indexed, but the book offers a brief bibliography of selected readings.

5-355 *Organic Guide to U.S. Colleges and Universities.* By the staffs of Environment Action Bulletin, Organic Gardening and Farming, and Fitness for Living. Emmaus, PA: Rodale, 1973. 213p.

The book reaches outside the purview of the bibliography to include information on organic gardening and establishing food cooperatives. Several chapters, however, are directly relevant, such as a chapter entitled "The Student as Environmental Activist" and a lengthy section on environment-related course offerings, which includes course descriptions as well as addresses for further information. Indexed.

5-356 Osborn, Fairfield. *Our Plundered Planet.* Boston: Little, Brown, 1948. 217p.

A book that details human consumption of the earth's irreplaceable natural resources. Written in a popular style, it contains chapters on Asia, Africa, the Soviet Union, Europe, and Australia. A lengthy bibliography concludes the volume. An important early book alerting the public to the "environmental crisis."

5-357 Osborn, Robert. *Mankind May Never Make It.* Greenwich, CT: New York Graphic Society, 1968. 92p.

This work is almost entirely pictorial, with line drawings and photographs expressing both the author's grim fears for the human race and his hopes that mankind *will* make it. Emphasis on human nature, the population increase, and destruction of natural resources.

5-358 Owen, Denis Frank. *Man in Tropical Africa: The Environmental Predicament.* New York: Oxford, 1973. 214p.

This book is about developmental dilemmas in Africa, where, according to the author, "some of the most rapid environmental changes that have ever occurred on earth are now taking place." The author investigates Africa's problems, some development proposals, and the probable biological consequences of those proposals. Chapters cover such topics as population, ecology of cultivation, weeds and pests, wild and domestic animals, and human diseases. Illustrated. Appended tables. Bibliography. Indexed.

5-359 Owen, Denis Frank. *What Is Ecology?* New York: Oxford, 1974. 188p.

A popular book that aims to "put across the essential scope of the subject and to place man in an ecological framework." Chapters on communities, food webs and organic diversity, ecosystems, and natural selection. A final chapter asks "How does ecology affect us?" Indexed.

5-360 Owen, Oliver S. *Natural Resource Conservation: An Ecological Approach.* New York: Macmillan, 1971. 593p.

Intended as a college text for the nonscience major, this book covers the history of the conservation movement, principles of ecology, water, rangelands, forest resources, wildlife, and fisheries. There are also sections on pesticides, air pollution, human population, and politics and citizen action. Name and subject indexes. Illustrated. Bibliography with each section.

5-361 Paradis, Adrian A. *Reclaiming the Earth: Jobs That Help Improve the Environment.* New York: McKay, 1971. 180p.

In this book descriptions of occupations are woven into a narrative of environmental concern. The author names real people working as city planners, architects, engineers, teachers, soil conservationists, public health officers, etc., and describes in some detail what they do in a typical day. Each chapter includes a list of organizations to write to for more information and suggested additional readings. Appended are an alphabetical list of professions keyed to literature available on each career, and a list of publishers' addresses. Indexed. Grades 7+.

5-362 Park, Robert Ezra. *Human Communities: The City and Human Ecology.* Glencoe, IL: Free Press, 1952. 278p.

This is volume 2 in the collected papers of Robert E. Park and consists of reprinted material on his pioneer studies of city life and an extensive section on human ecology. Among the papers reprinted in the second section are "Succession, an Ecological Concept," "Symbiosis and Socialization: A frame of Reference for the Study of Society," and "Human Ecology." Indexed by personal name and by subject.

5-363 Parnall, Peter. *The Mountain.* New York: Doubleday, 1971. 32p.

Illustrated by the author, this is a book about preserving the ecological balance and what happens when human beings interfere. The message is clear when a beautiful mountain becomes covered with trash. Grades K-3.

5-364 Paulsen, David F. and Denhardt, Robert B., eds. *Pollution and Public Policy: A Book of Readings.* New York: Dodd, Mead, 1973. 258p.

A collection of scholarly articles, the book is based upon the "interaction of technical and political processes in the formulation of environmental policy." The essays in part 1 outline the processes of policy-making in any field, while parts 2 and 3 deal with air pollution and water pollution policy-making, respectively. A two-paragraph précis of each article precedes the text of the essay. Intended for advanced students. Bibliography.

5-365 Perera, Thomas B. and Perera, Gretchen. *Louder and Louder: The Dangers of Noise Pollution.* New York: Watts, 1973. 42p.

This is a simple, clearly written little book which explains the nature of sound, when sound becomes noise, and the importance of controlling noise pollution. Well illustrated with black-and-white line drawings by Leonard Shortall. Indexed. Grades 2–4.

5-366 Perin, Constance. *With Man in Mind: An Interdisciplinary Prospectus for Environmental Design.* Cambridge: MIT Press, 1970. 185p.

The author hopes to bring together the social scientist and the environmental designer, and in her book she offers a "way of regularizing the infusion of the research capabilities and findings of the human sciences into the design process while it is going on." Appendix includes two lists of environmental attributes and resources considered by designers in the planning process. Bibliography. Indexed.

5-367 Perry, Bill. *Our Threatened Wildlife: An Ecological Study.* New York: Coward, McCann, 1969. 123p.

This book describes American species which are now extinct and those which have been saved from extinction by concerned citizens. The author also discusses the balance of nature and the need for wildlife management and research, with chapters on national parks, forests and refuges. Illustrated with black-and-white photographs. Indexed. Grades 7+.

5-368 Philip, Prince, Duke of Edinburgh and Fisher, James. *Wildlife Crisis.* New York: Cowles, 1970. 256p.

This book is really three books, so distinct are its three sections. The first part is Prince Philip's personal account of how his interest in bird watching grew into a more general concern for wildlife conservation. He also discusses techniques of wildlife photography, and includes his own excellent black-and-white photographs. Illustrated with many color plates, the second part of the book is a history of the conservation movement around the world, beginning in prehistoric times, by the naturalist James Fisher. The third section is an illustrated compendium of endangered species, showing which have vanished forever, which have been saved by human efforts, and which are currently threatened with extinction. Profusely illustrated. Selected bibliography. Indexed.

5-369 Pielou, E. C. *An Introduction to Mathematical Ecology.* New York: Wiley–Interscience, 1969. 286p.

An introduction to mathematical ecology with sections on population dynamics, spatial patterns in one-species populations, spatial relations of two or more species, and spatial relations in a many-species population. The book is technical in nature and indexed by author and subject.

5-370 Pinchot, Gifford. *Fight for Conservation.* Seattle: University of Washington Press, 1969. 152p.

A classic statement on the early perils of environmental neglect and the squandering of our natural resources. First published in 1910, it called for "special interests" to be put out of politics. Pinchot concluded, "I believe the young men will do it." Reprinted with an introduction by Gerald D. Nash.

5-371 Pirages, Dennis and Ehrlich, Paul R. *Ark II: Social Response to Environmental Imperatives.* New York: Viking, 1974. 344p.

A book that analyzes the politics of environmental problems, and also proposes social and political solutions. Social alternatives are presented as the prerequisite for environmental reform, and an agenda is set forth in rather complete detail. Bibliography. Indexed.

5-372 Platt, Robert B. and Griffiths, John F. *Environmental Measurement and Interpretation.* New York: Reinhold, 1964. 235p.

A text on instruments and the techniques of their utilization for measurement of the environment. Appendixes contain a guide to instrumentation and an extensive bibliography. Indexed.

5-373 *Politics of the Land: Ralph Nader's Study Group Report on Land Use in California.* Robert C. Fellmeth, project director. New York: Grossman, 1973. 715p.

This book, a lengthy volume in itself, is a condensation of *Power and Land in California*, a two-volume report prepared by the Center for Study of Responsive Law. Typical of the Nader reports, this is a well-documented study of land use and abuse in California. The text is divided into two sections, one on resources and one on development. Approximately one third of the volume consists of appended statistical material, such as tables of major private landowners and holdings in California, a listing of wildlife kills in 1969, information on State Water Plan finances, a list of state legislators' occupations, and so on. Notes collected at end of volume. Indexed.

5-374 *The Pollution Reader.* Compiled by Anthony DeVos, Norman Pearson, P. L. Silveston, and W. R. Drynan. Montreal: Harvest House, 1968. 264p.

This collection of readings results from a national Canadian conference, Pollution and Our Environment, organized and held late in 1966 by the Canadian Council of Resource Ministers. Essays on food, soil, water, air, and abatement and control. A list of contributors, a glossary, and a bibliography are contained in this book, which is not highly technical in nature. Not indexed.

5-375 Pomerantz, Charlotte. *The Day They Parachuted Cats on Borneo: A Drama of Ecology.* New York: Addison-Wesley, 1971. 64p.

A clever presentation, based on an actual event, of the problems human beings create when they interrupt natural cycles. In this case, DDT was used to kill mosquitoes in Borneo, setting off an unexpected and undesirable chain of events. Delightful illustrations. Grades 2–4.

5-376 *Power Generation and Environmental Change; symposium of the Committee on Environmental Alteration, American Association for the Advancement of Science, Dec., 28, 1969.* Edited by David A. Berkowitz and Arthur M. Squires. Cambridge: MIT Press, 1971. 440p.

A symposium on nuclear, hydroelectric and fossil-fuel power, with particular attention to waste heat, a product of both nuclear and fossil-fuel power plants. Additional papers were solicited for the book. Though the papers are technical in nature, most of the material is understandable by the layman. There are essays on the ecological effects of hydroelectric dams, nuclear reactors and the public health and safety, and environmental science and public policy. Amply illustrated. Indexed.

5-377 *Power, Pollution, and Public Policy; issues in electric power production, shoreline recreation, and air and water pollution facing New England and the nation.* Edited by Dennis W. Ducsik. Cambridge: MIT Press, 1971. MIT Report No. 24. 322p.

The results of a student Design Project for the spring semester, 1970. Chapters on offshore siting of electric power plants, shoreline recreation, controlling sulfur oxide emissions, water quality improvement in Boston harbor, and a prototype for regional government in New England. Indexed.

5-378 Priest, Joseph. *Problems of Our Physical Environment: Energy, Transportation, Pollution.* Reading, MA: Addison-Wesley, 1973. 389p.

Designed as a textbook for the college undergraduate, this book uses problems of the environment to illustrate principles of physics. It assumes only a basic knowledge of algebra. Each chapter concludes with questions for further discussion and exercises designated "easy to reasonable" or "reasonable to difficult." Appendixes include powers of ten and logarithms, and general properties of waves and thermal radiation. Glossary and index are included.

5-379 Pringle, Laurence P. *Ecology: Science of Survival.* New York: Macmillan, 1971. 152p.

A young-adult title which deals with the concept of ecology and animal populations. Extensively illustrated, it contains a glossary and a bibliography. Indexed.

5-380 Pringle, Laurence P. *One Earth, Many People: The Challenge of Human Population Growth.* New York: Macmillan, 1971. 86p.

The author approaches the problems and dilemmas of population control in a straightforward and understandable style. Attempting to present all points of view, the author does not back away from the topics of family planning, abortion, possible economic benefits for not having children, and so forth. Illustrated with black-and-white photographs. Glossary included. Indexed. Brief bibliography. Grades 7+.

5-381 Pringle, Laurence P. *Recycling Resources.* New York: Macmillan, 1974. 119p.

The author explains how politics interferes with recycling efforts. Both private industry and the federal government discourage recycling, according to Pringle. He is realisitc in saying that there is little the reader can do outside of writing letters, since major recycling must come from major industries and the industries are not yet attuned to the need for recycling. Grades 6–9.

5-382 *Radiation Hygiene Handbook.* Edited by Hanson Blatz. New York: McGraw-Hill, 1959. var. paging.

A manual by a number of contributors which details the scope and complexity of radiation hygiene. Chapters on nuclear safety, laboratory design, liquid and solid waste disposal, and control of radioactive air pollution. Includes reference data and a glossary of terms. Illustrated. Indexed.

5-383 *Project Survival*. Chicago: Playboy Press, 1971. 255p.

A collection of essays that were first published in *Playboy* magazine. Among the items reprinted is an interview with Paul Ehrlich and essays by Gene Marine, William O. Douglas, Loren Eiseley, and Sir Julian Huxley. Not indexed.

5-384 Purdom, P. Walton. *Environmental Health*. New York: Academic, 1971. 584p.

A text that provides a "comprehensive overview of man-environment-health interrelationships as well as a basic background for those working in any environmental health discipline." There are ten contributors to the text, in addition to the author, and the book contains chapters on food, water and waste water, air, solid waste, ionizing radiation, environmental control, the occupational environment, housing and the residential environment, accident prevention, and environmental planning and management. Indexed.

5-385 Randolph, Theron G. *Human Ecology and Susceptibility to the Chemical Environment*. Springfield, IL: Thomas, 1962. 148p.

Written by a doctor well over a decade ago, this book remains useful to the specialist and to the layman interested in the physical and mental effects on human beings of "mal-adaptation to the chemical environment." The author does not rail against air pollution or the introduction of dangerous chemicals into society—indeed, he ignores the political and social issues involved. This is a medical and scientific investigation which concerns itself with specific symptoms and treatments. Indexed.

5-386 Regenstein, Lewis. *The Politics of Extinction*. New York: Macmillan, 1975. 1,533p.

The first part of this book describes the depredations of hunters and the threat to ocean mammals and other wildlife. The second part is focused on the United States; the government's role in the extermination of several species, including the wild horse and the grizzly bear, is recounted in some detail. Introduction by Cleveland Amory.

5-387 Reid, George K. *Ecology of Inland Waters & Estuaries*. New York: Van Nostrand Reinhold, 1961. 375p.

A college text on aquatic ecology, this book is organized into five parts. It deals with the features and origins of basins and channels, the nature of water, physical and chemical characteristics of water, plants and animals which have adapted to inland waters, and population and community organization principles. Bibliography. Illustrated. Indexed.

5-388 *Remote Sensing in Ecology.* Edited by Philip L. Johnson. Athens: University of Georgia Press, 1969. 244p.

A series of technical papers on the use of computers, radioisotopes, and remote sensors in surveillance of the environment. Valuable as an insight into the range of technology now available to scientists. A lengthy and useful bibliography. Not indexed.

5-389 Resources for the Future. *Environmental Quality Analysis: Theory and Method in the Social Sciences.* Conference papers edited by Allen V. Kneese and Blair T. Bower. Baltimore: Johns Hopkins University Press (for Resources for the Future), 1972. 408p.

This book is organized around three main themes: understanding the relationship between the environment and economic growth, developing management programs, and designing political and legal institutions. Papers on agricultural pesticides, air pollution damage, and environmental quality as a problem of social choice. Since the published papers were revised in the light of discussions following their presentation, actual discussions of these papers are not reproduced in the book. Indexed.

5-390 Resources for the Future. *Environmental Quality in a Growing Economy; Essays from the Sixth Resources for the Future Forum.* By Kenneth E. Boulding and others. Edited by Henry Jarrett. Baltimore: Johns Hopkins University Press (for Resources for the Future), 1966. 173p.

Revised version of papers presented at the REF Forum on Environmental Quality held in March 1966. Compiled before "doomsday" literature began appearing in quantity, these papers are based on the assumption that the quality of life rather than survival of mankind is at issue in environmental problems. Six experts prepared papers on such topics as economics of the future, the environment and human health, side effects of resource use, and public attitudes on environmental quality.

Another expert in the field responded to each paper. The original statements and the responses to them make up the twelve essays of the book. Indexed.

5-391 Resources for the Future. *Perspectives on Conservation: Essays on America's Natural Resources.* By John Kenneth Galbraith and others. Baltimore: Johns Hopkins University Press (for Resources for the Future), 1969. 258p.

A forum, held in 1958, in which papers on a series of topics were presented over three months. The programs were "The First Fifty Years," "Science, Technology and Natural Resources," "Resource Demands and Living Standards," "Urban Growth and Natural Resources," "Some Determination of Resource Policy," and "Organizing for Conservation and Development." Under these programs, a variety of speakers offered essays from the professional fields of economics, geology, political science, geography, demography, public administration, and planning. Indexed.

5-392 Reynolds, William M. and Unger, James J. *Second Thoughts on Environmental Control.* Skokie, IL: National Textbook, 1970. 312p.

This book was developed from work done at the Georgetown University Forensic Institute in 1970, and it served as a casebook for the 1971 national high-school debates. Not indexed. (See also 6-11.)

5-393 Ridgeway, James. *Last Play: The Struggle to Monopolize the World's Energy Resources.* New York: Dutton, 1973. 446p.

An examination of the financial control of the world's energy resources, with chapters on oil, natural gas, uranium, coal, transportation in the United States, future fuels, and major energy corporations. An exhaustive chart on financial institutions' control of energy companies completes the book. Indexed.

5-394 Ridgeway, James. *The Politics of Ecology.* New York: Dutton, 1970. 222p.

An important inquiry into the political conditions which govern environmental pollution and the prospects for remedy. Ridgeway's conclusion suggests that "stripped of its façade, Environmental Action could offer no more than any other new industry, whose growth was tied to increasing pollution." The politics and historical significance of sewage and oil pollution are dealt with in this volume. It carries a brief section on sources, and advances some general recommendations towards alleviating pollution— by changing our national fuel policy, for example. Indexed.

5-395 Rienow, Robert and Rienow, Leona Train. *Moment in the Sun: A Report on the Deteriorating Quality of the American Environment.* New York: Dial, 1967. 286p.

A status report on the quality of our environment, which involved four years of preparation and twenty-five thousand miles of travel by the authors. Extensive chapter notes are found at the end of the book. An excellent survey of informal opinion about the environmental crisis at the time this book was written. Indexed.

5-396 Riley, Herbert P. *Evolutionary Ecology.* Belmont, CA: Dickenson, 1970. 113p.

The book is designed for upper-level college courses in ecology, and it offers readings on the nature of variation, natural selection, ecological adaption, and ecological evolution. The material is technical in nature, and bibliographies of literature cited accompany each reprinted article. Indexed.

5-397 Robinson, John. *Highways and Our Environment.* New York: McGraw-Hill, 1971. 352p.

A complete examination of the highway in our society. Chapters on "The Road in Our Civilization," "Visual Excrescence and other Evils," "Bringing Back the Stile," and "The Political Arena." An appendix contains a bibliography, extensive notes on the highway lobby, a list of citizen action groups, and sources of funds for scenic improvement of streets and highways. Bountifully illustrated. Indexed.

5-398 Rockefeller, Nelson A. *Our Environment Can Be Saved.* Garden City, NY: Doubleday, 1970. 176p.

A personal account of environmental protection programs undertaken by the author when he was governor of New York, with many references to federal programs and problems in other states. The book, optimistic in outlook, is based on the assumption that economic growth and environmental protection are not mutually exclusive. The final chapter offers guidelines for citizen action. Appended are lists of federal and state agencies in the environmental field, relevant private organizations, and a brief annotated bibliography.

5-399 Rodgers, C. Leland and Kerstetter, Rex E. *The Ecosphere: Organisms, Habitats, and Disturbances.* New York: Harper, 1974. 341p.

The purpose of this college text is to "assess man's impact and potential in the natural environments," and to consider environmental problems and possible solutions. Illustrated amply, it contains a glossary and a list of parks and nature preserves, with a brief description of each area. Indexed.

5-400 Rodgers, William. *Brown-Out: The Power Crisis in America.* New York: Stein & Day, 1972. 300p.

A full examination of the problems of electric power in the United States and alternative power sources. Written for a general audience, this book provides a historical account of recent criticism of the AEC and deals extensively with the problems of nuclear power. Indexed.

5-401 Rondiere, Pierre. *Purity or Pollution: The Struggle for Water.* New York: Watts, 1971. 128p.

The author begins with brief scenarios of water-related crises developing around the world. He then discusses the essential need for pure water, the scientific nature of water, how organisms use water, the way water is overloaded with waste, the world's thirst, and how water can be conserved and purified. Illustrated with many photographs, some in color. Indexed. Grades 6-9.

5-402 Roos, Leslie L. *The Politics of Ecosuicide.* New York: Holt, 1971. 404p.

A book of readings prepared for college courses in political science and "interdepartmental offerings concerning the environment." Some essays were originally written for this book. Sections on public opinion, new institutions, bureaucracy, the role of analysis, and the problems of change. An appendix contains a suggested course outline on politics and the environment. and includes a bibliography. Not indexed.

5-403 Roosevelt, Nicholas. *Conservation: Now or Never.* New York: Dodd, Mead, 1970. 238p.

The author, a cousin of Theodore Roosevelt, is concerned with "saving scenic resources." The book offers a historical perspective on the changes which have occurred in preservation of natural resources. Case histories are used to illustrate opposing forces in the conservation struggle. Excludes water and air pollution per se; focuses on preservation of irreplaceable land areas. Indexed.

5-404 Rosenbaum, Walter A. *The Politics of Environmental Concern.* New York: Praeger, 1973. 298p.

Using case studies of ecological disasters, the author documents various threats to the environment. He discusses existing legislation and the adversaries involved in the fight to determine who will control the environment. Chapter 3 describes in some detail the major organizations concerned with protection of the environment. Other chapters investigate the politics of timber, water resources, regulating air and water pollution, surface mining, and solid waste disposal. Indexed.

5-405 Rosenkrantz, Barbara Gutmann and Koelsch, William A., eds. *American Habitat: A Historical Perspective.* New York: Free Press, 1973. 372p.

Two history professors present a broad range of contributed essays based on the premise that "man's habitat is constructed out of his image of society as much as his image of nature, that his use of land and water reflect a complex of political, social and ideological factors. . . ." The book is divided into four sections: The first deals with questions of public policy and the environment, the second traces the history of settlement in the United States, the third analyzes various nineteenth-century paintings, and the fourth discusses questions of environmental values. Illustrated. Indexed.

5-406 Ross, Richard D. *Air Pollution and Industry.* New York: Van Nostrand Reinhold, 1972. 489p.

Intended as a handbook for plant engineers, this book is written for business and industry. It offers comprehensive coverage of the problems of air pollution and its control, with sections on analytical techniques, the design of a plant air management system, selection of equipment for particulate removal, and so on. There is also a major section on legislative and regulatory trends in air pollution control, with which industrial engineers should be familiar. Each section includes a list of references. Illustrated. Indexed.

5-407 Rothman, Harry. *Murderous Providence: A Study of Pollution in Industrial Societies.* Indianapolis, IN: Bobbs-Merrill, 1972. 372p.

The author argues that "the private economic rationality of the profit-seeking business enterprise is a murderous providence because it cannot guarantee the optimum use of resources . . . [and] it cannot avoid continually creating situations which cause the pollution of our environment." Part 2 of the book consists of a survey of the worldwide environmental crisis, while part 3 is devoted to the techniques of pollution control. Extensive notes appear in a separate section, and the book contains a long and useful bibliography. Indexed.

5-408 Roueche, Berton. *What's Left: Reports on a Diminishing America.* Boston: Little, Brown, 1962. 210p.

A collection of essays by a naturalist who is also a writer. Each chapter details an experience in nature in well-turned prose. The overall impact of the book is to point out what has been lost and what remains of nature for people to enjoy.

5-409 Rubin, David M. and Sachs, David P. *Mass Media and the Environment: Water Resources, Land Use and Atomic Energy in California.* New York: Praeger, 1973. 319p.

An attempt to investigate the role of the media in the public's understanding of the environmental crisis. Most of the research applies to the San Francisco Bay Area, but it is nonetheless of interest on a national level. Chapters on environmental reporters, the environmental information explosion in mass market publications, and other topics. Bibliography, Indexed.

5-410 Rudd, Robert L. *Pesticides and the Living Environment.* Madison: University of Wisconsin Press, 1964. 320p.

An exhaustive analysis of the effects of pesticides upon the environment. Useful both as an introduction to pesticides in use and to the legislative control of them. An appendix contains *Recommendations of the President's Scientific Advisory Committee on Use of Pesticides* and a long list of references. Indexed.

5-411 Russell, Helen Ross. *Earth, the Great Recycler.* Nashville: Nelson, 1973. 160p.

The author explains the basic elements of the earth's natural systems, the perpetual motion of water and of energy, the food chain, etc. An understanding of life's cycles is the basis for the author's urgent plea that man "fit his activities into the pattern." Problems of pesticides, mercury poisoning, and energy consumption are discussed. The author recommends action that young people can undertake to help protect the environment. Endpapers include a map of earth's environmental problems. Indexed. Grades 7+.

5-412 Saltonstall, Richard. *Your Environment and What You Can Do About It.* New York: Walker, 1970. 299p.

This "prescription for effective citizen action" includes basic information on water pollution, air pollution, noise pollution, problems of growth, solid waste, pollution of shorelines, and land use, with separate chapters on the environmental laws and a checklist for individuals who want to conserve resources. The author reports throughout on successful efforts by citizens and citizen groups to solve environmental problems. Appended material includes a list of agencies and organizations, key legislative committees, and a bibliography. Indexed.

5-413 Salvato, Joseph A. *Environmental Engineering and Sanitation.* 2d ed. New York: Wiley–Interscience, 1972. 919p.

A textbook that deals with diseases, water supply, waste water treatment and disposal, solid waste management, air pollution, radiation, food, recreation areas, weed control, housing, and the residential environment. Illustrated. Indexed.

5-414 Sax, Joseph L. *Defending the Environment: A Strategy for Citizen Action.* New York: Knopf, 1971. 252p.

The author, a law professor at the University of Michigan, provides the reader with an in-depth perspective on the "mind-set" of the law and its officers. This is not simply an outline or list of things to do, but rather an analysis of how the judicial system works, which the author hopes will encourage greater citizen participation. Case histories are used to explain the process. The book includes an introduction by Senator George McGovern. Appended is a model environmental law drafted by the author and introduced in the Michigan Legislature in 1969. Indexed.

5-415 Scheffer, Victor B. *A Voice for Wildlife.* New York: Scribner, 1974. 245p.

This book is intended to "stimulate thought about wildlife

management and to predict the direction in which it is moving."
The material in the book is developed under three major topics:
confrontation between people and wildlife, wildlife management
and how it works, and an ethic toward wildlife. The author,
formerly a biologist in the Forest Service, National Park Service,
Fish and Wildlife Service, and National Marine Fisheries Service, is interested in attitudes toward wildlife management.
Notes are collected at the end of the volume. Brief bibliography.
Indexed. Illustrated.

5-416 Schlichting, Harold E. and Schlichting, Mary Southworth.
Ecology: The Study of Environment. Austin, TX: Steck-Vaughn,
1971. 48p.

This brief, amply illustrated text is a primer in ecology. The
authors define and present, in simple language, concepts such as
ecosystem, food chain, and competition and cooperation among
plants and animals. The topic of pollution is introduced in a
chapter on the human effect upon the environment. Well illustrated with line drawings by Don Collins. Grades 3–5.

5-417 Schneider, Kenneth R. *Autokind vs. Mankind.* New York:
Norton, 1971. 267p.

The author, a city planner, considers in some detail the automobile's influence on American life. Having come to the conclusion that the automobile is a destructive force in our lives, the
author outlines the history and development of the auto and describes how it now threatens our environment, our mental health,
and our society. He advocates ways to "recapture" the country
from "autokind."

5-418 Schroeder, Henry Alfred. *The Poisons Around Us: Toxic
Metals in Food, Air and Water.* Bloomington: Indiana University
Press, 1974. 144p.

This book discusses metals and other elements as pollutants
and as beneficial substances. The author, a doctor primarily
concerned with human health, concentrates on lead, mercury,
and cadmium, but includes material on other elements either
nontoxic or beneficial to humans. Brief bibliography. Indexed.

5-419 Schwartz, George I. and Schwartz, Bernice S. *Food Chains
and Ecosystems: Ecology for Young Experimenters.* Garden City,
NY: Doubleday, 1974. 109p.

The authors define ecology, distinguish between research by

observation and experimentation, give safety warnings about harmful flora and fauna, list necessary equipment for suggested experiments, and discuss interdependence and chains in nature. Bibliography included. Indexed. Grades 6-9.

5-420 Schwartz, William, ed. *Voices for the Wilderness*. New York: Ballantine, 1969. 366p.

A selection of more than thirty articles from Sierra Club publications, this book surveys the problems of preserving the wilderness, the value of the wilderness, and programs for preservation in the future. Authors include Stewart Udall, William O. Douglas, and Ashley Montagu.

5-421 Scorer, Richard S. *Pollution in the Air: Problems, Policies, and Priorities*. London: Routledge & Kegan Paul, 1973. 148p.

Dealing with worldwide problems of air pollution, this British author describes the scientific nature of the atmosphere, how the air is naturally cleaned, and the need for political and economic changes which will provide for further cleaning of the air. Examples are chosen with Britain in mind and the section on antipollution law may be irrelevant to the American reader, but on the whole this is a useful text. Indexed.

5-422 Sears, Paul B. *Lands Beyond the Forest*. Englewood Cliffs, NJ: Prentice-Hall, 1968. 206p.

A book on the natural history of grasslands, their evolution, and their continuing importance in human life. Poetically written. Illustrated. Indexed.

5-423 Sears, Paul B. *This Is Our World*. rev. ed. Norman: University of Oklahoma Press, 1971. 294p.

This book was first published in 1937 and was "an informal experiment . . . to combine . . . the wisdom to be derived from both cultural anthropology and ecology." In the revised edition, the last chapter has been rewritten, and the original text changed to reflect the passage of time. Illustrated. Not indexed. An important book in a popular style about the wisdom of relating properly to our environment.

5-424 Sears, Paul B. *Where There Is Life*. New York: Dell, 1970. 205p.

An introduction to ecology, written for the general reader, that has chapters on the formation of ecology as a discipline and a great deal of American natural history. While it is somewhat dated in terms of contemporary developments, many readers will still find it a useful book. Indexed.

5-425 Segerberg, Osborn. *Where Have All the Flowers, Fishes, Birds, Trees, Water, and Air Gone?* New York: McKay, 1970. 303p.

The author, a journalist, defines and documents the environmental crisis. Part 1 details "the system" of nature, showing how interrelated all species are. Part 2 chronicles human interruption of the system and the results of that disruption. Heavily documented, but written for the layman. Indexed.

5-426 Shanks, Ann Zane. *About Garbage and Stuff.* New York: Viking, 1973. unpaged.

The author uses a family's normal practices in shopping for groceries and throwing away garbage as a framework for a discussion of recycling. The family learns to separate glass, metal, and paper from garbage and the reader learns how recycling plants work. Illustrated with large photographs by the author. Grades 3-5.

5-427 Shelford, Victor E. *The Ecology of North America.* Urbana: University of Illinois Press, 1963. 610p.

This "ecological reconstruction" of North America during the sixteenth century provides in-depth descriptions of biotic communities native to the area prior to European settlement. The first chapter on the scope and meaning of ecology serves not only as an introduction but also as a glossary. The rest of the volume is organized by type of community, such as temperate deciduous forest biome, with each section providing information on characteristic species, populations of animals and densities of plants, food habits of animals, and interrelationships between animals and vegetation. Includes bibliography of all literature cited, locality index, and species index.

5-428 Shepard, Paul and McKinley, Daniel, eds. *Environmental Essays on the Planet as a Home.* Boston: Houghton Mifflin, 1971. 308p.

A collection of readings on human ecology intended for the college classroom or the interested layman. Attention is paid to emotional and esthetic issues, the social and psychological impact of overpopulation, and specific problems such as air pollution. Each essay includes its own list of references. The book concludes with biographical notes on the contributors and an extensive bibliography of additional readings. Illustrated.

5-429 Shepard, Paul and McKinley, Daniel, eds. *The Subversive Science: Essays Toward an Ecology of Man*. Boston: Houghton Mifflin, 1969. 453p.

A group of readings, designed for the college classroom, that brings broad social concerns to bear on problems stemming from human relationships to the environment. Sections on human beings as populations, human beings and other organisms, human beings in ecosystems, and ethos, ecos, and ethics. A long bibliography of additional readings is included. Illustrated.

5-430 Shomon, Joseph James. *Open Land for Urban America: Acquisition, Safekeeping, and Use*. Baltimore: Johns Hopkins University Press, 1971. 171p.

A book that argues for the preservation of open space in the urban landscape. Illustrated with photographs and maps; appendixes carry samples of legal documents for acquiring scenic easements in Wisconsin and New York. Bibliography. Indexed.

5-431 Shuttlesworth, Dorothy Edwards. *Clean Air, Sparkling Water*. Garden City, NY: Doubleday, 1968. 95p.

The author discusses the causes and effects of air and water pollution, using the experiences of a hypothetical city and nearby town. Also included is a perspective on pollution in Europe and Asia. The book concludes with a discussion of what is being done to clean up the air and water and what still needs to be done. Illustrated with black-and-white photographs. Indexed. Grades 3-5.

5-432 Silverstein, Alvin and Silverstein, Virginia. *The Chemicals We Eat and Drink*. Chicago: Follett, 1973. 112p.

The authors discuss natural and artificial poisons in food, the beneficial effects of some chemicals, foods of the future, diet control, and pollution control. Indexed. Grades 5-8.

5-433 Simon, Noel and Geroudet, Paul. *Last Survivors: The Natural History of Animals in Danger of Extinction*. New York: World, 1970. 275p.

This book relates the natural history of selected endangered animals—thirty-six mammals and twelve birds. Two or three pages of text are given to each species, and each animal is portrayed either in a full-color painting or a black-and-white line drawing contributed by Helmut Diller or Paul Barruel, two artists who are also naturalists. Prepared under the auspices of the World Wildlife Fund. Bibliography.

5-434 Sloan, Irving. *Environment and the Law*. Dobbs Ferry, NY: Oceana, 1971. 120p.

This small book, number 65 in the Legal Almanac Series, outlines environmental law in nontechnical language for the general reader. Chapters treat the structure of federal environment control, air pollution, water pollution, noise pollution, pesticides, land use, procedural aspects of environmental law, international regulation for environmental quality, and an analysis of antipollution legislation and case law. Appendixes include several key pieces of national environmental legislation, a list of conservation periodicals, and a bibliography. Brief index.

5-435 Small, William E. *Third Pollution: The National Problem of Solid Waste Disposal*. New York: Praeger, 1971. 173p.

The author reviews the progress that has been made in product recycling, improved collection techniques, and more effective means of disposal. His message, however, is that much more needs to be done. Separate chapters on agricultural waste, mineral resources and energy wastes, and urban waste. Bibliography. Indexed.

5-436 Smith, Grahame J. C.; Steck, Henry J.; and Surette, Gerald. *Our Ecological Crisis: Its Biological, Economic, and Political Dimensions*. New York: Macmillan, 1974. 198p.

The authors, professors of biology, political science, and economics respectively, present an integrated argument for the development of an environmental ethic. The first section offers an ecological framework for the rest of the text. The second deals with the cultural and economic bases of the crises, and the final section outlines the involvement of the federal government in the environmental crisis. Economic growth—"so necessary to our economic health, so deadly to the natural world"—is the topic of an epilogue. Each section includes suggested readings. Illustrated. Indexed.

5-437 Smith, Guy-Harold, ed. *Conservation of Natural Resources*. 4th ed. New York: Wiley, 1971. 685p.

A general textbook on natural resources, this book includes

sections on conservation, soil and pollution, water, minerals, forest, wildlife, recreational uses of resources, and planning for the future. Most chapters are contributed by other specialists; however, the editor is author of several articles. Illustrated. Bibliography. Indexed.

5-438 Smith, Robert L., ed. *The Ecology of Man: An Ecosystem Approach.* New York: Harper, 1972. 436p.

An overview of human ecology, this is a collection of essays supplemented by the editor's own writing. The book includes sections on the concepts of the ecosystem and the food chain, the human habitat, population, endangered environments, and the future. Illustrated. Selected references given at end of volume. Indexed.

5-439 Sollers, Allan A. *Ours Is the Earth: Appraising Natural Resources and Conservation.* New York: Holt, 1963. 128p.

A book which stresses the interdependence of our basic resources and discusses water, soils, forests, rangeland, wildlife, and mammals. Present and future conservation practices are discussed and appraised. Indexed, with a short glossary.

5-440 Sootin, Harry. *Easy Experiments With Water Pollution.* New York: Four Winds, 1974. 109p.

Well-written and thorough, this book presents discussions on chlorination, biodegradation, and septic tank processes, among other topics. Experiments are clearly described for each major topic; sophisticated, expensive equipment is not needed for most experiments. Glossary and bibliography included. Grades 5-7.

5-441 Southwick, Charles H. *Ecology and the Quality of Our Environment.* New York: Van Nostrand Reinhold, 1972. 319p.

A textbook written for college students who are not majoring in ecology, this volume could easily be used by the layman for background information. The text begins (rather than ends) with a section on environmental deterioration and how it affects man. Subsequent chapters describe the history of ecology and pollution, basic ecological principles, population ecology, and community ecology. Bibliography. Glossary. Illustrated. Indexed.

5-442 Spedding, C. R. W. *Grassland Ecology*. Oxford, Eng.: Oxford University Press, 1971. 221p.

Chapters on the individual plant, plant populations, the fauna of grasslands, grass as food for animals, and animal production from grasslands. The book is technical in nature and attempts to define a model for use in grassland management. Indexed by author and subject.

5-443 Sprout, Harold and Sprout, Margaret. *The Ecological Perspective on Human Affairs, With Special Reference to International Politics*. Princeton, NJ: Princeton University Press, 1965. 236p.

This book represents a further development of the authors' earlier essay, "Man-Milieu Relationship Hypotheses in the Context of International Politics" (1956). The previous work attempted "to bring the disciplines of geography and politics into more productive relationships." The present book argues that determinism of any sort or "the practice of explaining past events and predicting future trends in the patterns of international politics by some reference to some set of environmental factors" does not in any way aid our understanding in international politics. Indexed.

5-444 Spurr, Stephen Hopkins and Barnes, Burton V. *Forest Ecology*. 2d ed. New York: Ronald, 1973. 571p.

This textbook deals with the ecological basis for the management of forested land. It is divided into four sections: The forest tree, the forest environment, the forest community and ecosystem, and the forest as a whole. Each chapter contains a summary and suggested further readings. The book is illustrated and contains a list of the scientific names of trees and an extensive bibliography. Indexed.

5-445 Stacks, John F. *Stripping*. San Francisco: Sierra Club, 1972. 140p.

This brief monograph focuses on the surface mining of coal in America. It is a cry against the damage done to the environment and the people by strip-mining processes. The author discusses the technology involved in taking coal from the earth, the environmental and social impact of the process, regulations and reclamation procedures, the power of vested economic interests, and fuel conglomerates. The author concludes with the hope that a new modesty is replacing the old American arrogance toward the natural systems on which life depends. Not indexed.

5-446 Steinhart, Carol E. and Steinhart, John S. *Blowout: A Case Study of the Santa Barbara Oil Spill.* North Scituate, MA: Duxbury, 1972. 138p.

A detailed study of the political, ecological, and scientific results of a major disaster. In addition to recounting the events during and after the spill, the authors offer historical perspectives on the geologic and economic development of the area, the techniques of oil drilling, and government leasing. Finding no individual or corporate villain to bear the full blame for the Santa Barbara spill, or full responsibility for future disasters, the authors point to this country's "energy addiction" as the underlying cause. Map of Santa Barbara area and chronology of events included.

5-447 Stern, Arthur Cecil. *Air Pollution.* 2d ed. New York: Academic, 1968. 3 vols. 694p., 684p., 868p.

This three-volume work is concerned with the cause, effect, movement, and control of air pollution. It is written for the "scientifically advanced reader," with each chapter contributed by a different author. Volume 1 covers the nature of air pollution, how it spreads, and its effects. Volume 2 includes material on sampling, analysis, measurement, and monitoring of air pollution. Volume 3 covers emissions from principal air pollution sources; control techniques and equipment; laws, regulations and standards; and administrative procedures used to enforce those laws. The final chapter of Volume 3 is a discussion of other sources of information about air pollution. Each volume includes an author and subject index.

5-448 Stevens, Leonard A. *The Town That Launders Its Water: How a California Town Learned to Reclaim and Reuse Its Water.* New York: Coward, McCann, 1971. 122p.

A case study written for younger readers about an experiment in water reuse in Santee, California. The author begins by describing the problem—a drop in the water table. Then he tells how the reuse project was funded, a purification plant constructed, and public acceptance gained for cleaning and reusing sewage water. The town now has eight man-made lakes. Black-and-white photographs. Indexed. Grades 6-8.

5-449 Stewart, George Rippey. *Not So Rich As You Think.* Boston: Houghton Mifflin, 1967. 148p.

A popular monograph on the problems of waste, this book discusses waste disposal in down-to-earth language. The serious problems of sewage, factory effluents, garbage, litter, mineral and agricultural refuse, smoke, and smog are lightened by the author's sense of humor, as well as by Robert Osborn's line drawings. Illustrated. Indexed.

5-450 Still, Henry. *In Quest of Quiet: Meeting the Menace of Noise Pollution—Call to Citizen Action.* Harrisburg, PA: Stackpole, 1970. 221p.

Written for the layman, this is a survey of noise pollution and its effects, both psychological and physiological. The author identifies sources of noise pollution, reports results of research, and recommends in the final chapter courses of action open to citizens. References are collected at the end of the book. Not indexed.

5-451 Still, Henry. *Man: The Next 30 Years.* New York: Hawthorn, 1968. 216p.

The author looks into the immediate future (which is now) and predicts the state of food production, water, energy, transportation, automation, education, and so on. While noting humanity's past errors, the author is optimistic about the future. His predictions are really statements of potential, describing what man has the knowledge and power to do if he has also the will. A book of documented dreams. Indexed.

5-452 Stoker, H. Stephen and Seager, Spencer L. *Environmental Chemisty: Air and Water Pollution.* Glenview, IL: Scott, Foresman, 1972. 186p.

A college textbook on the chemistry of air and water pollution. Supplementary readings are recommended at the end of each chapter. Indexed.

5-453 Strobbe, Maurice A. *Understanding Environmental Pollution.* St. Louis: Mosby, 1971. 357p.

This collection of articles is intended as a supplementary text for college courses, both in the sciences and the humanities. In order to serve both, the articles have been divided into two parts.

First, general articles written in nontechnical language cover such topics as water pollution, DDT, and population control. The second group of articles includes research reports and technical investigations into the problems outlined in part 1. Appendixes include a brief bibliography, a list of selected films on environmental issues, a list of national conservation organizations, and suggested laboratory tests and demonstrations.

5-454 Study of Man's Impact on Climate, Stockholm, 1970. *Inadvertent Climate Modification: Report.* Cambridge: MIT Press, 1971. 308p.

This project "reviewed SCEP findings critically [see 1-6], to point to global environmental problems that were . . . overlooked; to obtain a more complete assessment of present knowledge . . . and to point to questions requiring international policy decisions." Parts 1 and 2 provide a view of the topic for the lay reader while parts 3 and 4 present detailed assessments of knowledge about specific areas. Chapter 2 presents the major recommendations and conclusions of the study. Indexed.

5-455 Symposium on the Global Effects of Environmental Pollution, 1st, Dallas, TX, 1968. *Global Effects of Environmental Pollution.* Edited by S. Fred Singer. New York: Springer, 1970. 218p.

Technical papers devoted to chemical balance of gases in the earth's atmosphere, nitrogen compounds in soil water, atmosphere and precipitation, effects of atmospheric pollution on climate, and worldwide ocean pollution by toxic water. Illustrated. Biographical notes on authors. Indexed.

5-456 *Systems Analysis and Simulation in Ecology.* Edited by Bernard C. Patten. New York: Academic, 1971–1972. 2 vols. 607p., 592p.

The idea of this book was to "draw together a collection of systems ecology articles as a convenient benchmark to the state of this new emerging field." Computer simulation models are presented and discussed for a variety of ecosystems. Highly technical and mathematical in nature, this book was written by a number of contributors. Indexed.

5-457 Talbot, Allan R. *Power Along the Hudson: The Storm King Case and the Birth of Environmentalism.* New York: Dutton, 1972. 244p.

This is a detailed case history of the Storm King case, in which the courts held for the first time that citizens with no economic interest in the property under dispute have a legitimate role in decisions regarding use of that property. Chapters on the natural history of Storm King Mountain, Con Ed, New York City's power needs, the trial, and its aftermath. Readable and balanced account of a landmark case. Acknowledgments and bibliography for each chapter gathered at the end of the volume. Indexed.

5-458 Tamplin, Arthur R. and Gofman, John W. *Population Control Through Nuclear Pollution*. Chicago: Nelson-Hall, 1970. 242p.

The examination of the danger and hazards of nuclear programs sponsored by the Atomic Energy Commission. The authors, research associates at the Lawrence Radiation Laboratory, provide an introduction to the perils of low-dose radiation hazards. The book contains a foreword by Paul Ehrlich. Not indexed.

5-459 Taylor, Gordon Rattray. *The Doomsday Book: Can the World Survive?* New York: World, 1970. 335p.

Written by a British journalist with formal training as a biologist, this book explores many threats to the environment and to human survival on earth. Documented with numerous references to scholarly studies, it is written in a popular, readable style. A list of organizations concerned with the environment is contained in an appendix.

5-460 Taylor, Theodore B. and Humpstone, Charles C. *The Restoration of the Earth*. New York: Harper, 1973. 166p.

A book by a nuclear physicist and a lawyer that is "less concerned about pollution problems than how to go about solving them." The authors advance the "containment principle"—the requirement that the environmental effects of human activity be confined within areas dedicated to that activity. Its application dictates that all substances and forms of energy must be stored or recycled within defined portions of the earth's surface. A final chapter is devoted to the international ramifications of pursuing this policy. Not indexed.

5-461 Teal, John and Teal, Mildred. *Life & Death of the Salt Marsh*. Boston: Little, Brown, 1969. 278p.

A book on the ecology of the salt marsh. Illustrated by Richard Fish, it supplies a pictorial as well as narrative account of the marshes of the east coast of North America. It explains how they were formed, why it is they continue, and the effect of their presence on man and vice versa. A final chapter offers suggestions to prevent the systematic destruction of these ecosystems. Indexed.

5-462 Terry, Mark. *Teaching for Survival.* New York: Ballantine, 1971. 213p.

A basic text on the importance, nature, and role of environmental education. There is a long section on the function of the school, and a useful bibliography offers education plans for material covered and, by extension, for any course taught utilizing the recommended material.

5-463 Thomlinson, Ralph. *Urban Structure: The Social and Spatial Character of Cities.* New York: Random House, 1969. 335p.

This book focuses on what the author calls "urban ecology," which he defines as "human spatial distribution in and around cities as an influence and mediator of social relations." Part 1 of the book is a history and sociological analysis of cities; part 2 is an analysis of various theories of urban ecology; and part 3 is devoted to city planning and possibilities for the future. The volume concludes with selected readings, a name index, and a subject index. Charts and graphs included.

5-464 Thompson, Donald N. *The Economics of Environmental Protection.* Cambridge, MA: Winthrop, 1973. 278p.

The book is "an economist's view of environmental protection" and is not intended as a "pedantic application of microeconomic theory and welfare economics to pollution problems." Chapters on air pollution, water pollution, chemical pollution, and on both economic and legal approaches to cost internalization. Two chapters on the National Environmental Policy Act and The Stockholm Conference complete the book. Indexed, with a glossary.

5-465 Treshow, Michael. *Whatever Happened to Fresh Air?* Salt Lake City: University of Utah Press, 1971. 201p.

The author of this book is a biologist. Written in a popular style, the book deals with all aspects of air pollution, including the threat to human beings and the environment. The book is not indexed and does not carry recommendations for further reading, but it is complete in scope and thorough in its treatment of the subject.

5-466 Troost, Cornelius J. and Altman, Harold, eds. *Environmental Education*. New York: Wiley, 1972. 575p.

The book addresses two major questions: What is the nature of the environmental crisis? and What learning experiences should be provided by this country's schools to meet this crisis? Part 3 offers an integrated ecology curriculum from kindergarten through high school. Specific recommendations are made for schools in urban areas. Indexed.

5-467 Turk, Amos; Turk, Jonathan; and Wittes, Janet T. *Ecology, Pollution and Environment*. Philadelphia: Saunders, 1972. 217p.

A textbook on ecology and the environment that deals in separate sections with agricultural environments, pesticides, radioactive wastes, air pollution, water pollution, solid wastes, growth of human populations, thermal pollution, and noise. An appendix includes a section on the metric system, chemical symbols, formulas, and a table of relative atomic weights. Well illustrated, each chapter concludes with selected further reading and a list of discussion questions. Indexed.

5-468 Turner, James S. *The Chemical Feast*. New York: Grossman, 1970. 273p.

A study of the functioning of the U.S. Food and Drug Administration by a Ralph Nader Study Group. Food additives are considered in depth, as is the failure of this federal agency to represent the public interest in protecting the health of the nation. The study was conducted and completed in 1969. Indexed.

5-469 Udall, Stewart L. *1976: Agenda for Tomorrow*. New York: Harcourt, 1968. 173p.

In a particularly lucid foreword, former Secretary of the Interior Stewart Udall notes that "the total-environment approach of 'the new conservation' demand[s] concepts large enough to relate conservation to the overriding issues of our age." Thus this book deals not only with air and water pollution, but with troubled cities, the quality of life, and the renewal of politics. An optimistic search for an agenda for tomorrow. Annotated list of books, organized by chapter. Indexed.

5-470 Ullrich, Wolfgang. *Endangered Species*. New York: Hart, 1971. 284p.

The author, longtime director of the Dresden Zoo, offers descriptions of 109 birds and mammals considered scarce or endangered. The text, approximately one page for each species, is accompanied by black-and-white and color photographs of the animals and birds. The author describes hunting methods, natural enemies, habits, size of litter, and so forth. Not indexed.

5-471 *Vanishing Air: The Ralph Nader Study Group Report on Air Pollution.* John C. Eposito, project director; Larry J. Silverman, associate director. New York: Grossman, 1970. 328p.

A study growing out of the 1969 Nader project to examine the National Air Pollution Control Administration. Air pollution enforcement is studied throughout the country, as well as on the federal level, in this report. Indexed.

5-472 Vayda, Andrew P. *Environment and Cultural Behavior: Ecological Studies in Cultural Anthropology.* Garden City, NY: Natural History, 1969. 485p.

A group of twenty-three essays by cultural anthropologists on relating "cultural behavior to environmental phenomena." All essays except two have been previously published. Essays are grouped under two topics: systems in operation, and origins and development. Indexed.

5-473 Vogt, William. *Road to Survival.* New York: Sloane, 1948. 335p.

An early and rather thorough treatment, from the standpoint of human ecology, of the problems of mankind's survival. Attention is paid to all parts of the world in this book. It contains an introduction by Bernard M. Baruch, a section for author notes, and recommendations for further reading. Exhaustively indexed.

5-474 Wagner, Philip Laurence. *The Human Use of the Earth.* Glencoe, IL: Free Press, 1960. 270p.

According to the author, the themes of this book are both geographical and ecological. Chapters on human societies as geographic forms, the means of production and the commercial environment; final chapter on a geographic outlook. The book contains an extensive section of notes. Indexed.

5-475 Wagner, Richard H. *Environment and Man.* New York: Norton, 1971. 491p.

Written by a botanist, this book is designed to provide an

introductory course for college students in ecology, i.e., biology. Chapters on biocides, inorganic pollutants, and the introduction of exotics into our ecosystems make it a relevant text. A section on the urban environment informative. Includes photographs, selected suggested readings, and an index.

5-476 Walker, Laurence C. *Ecology and Our Forests*. South Brunswick, NJ: A. S. Barnes, 1972. 175p.

The author, dean of the school of forestry at Austin State University, outlines the interrelationships occurring in forests, discussing in detail the characteristics and requirements of some twenty types of trees indigenous to North America. The discussion includes a concept of forest management which treats forests as renewable resources to be harvested and used. Brief section of black-and-white photographs. Glossary. Indexed.

5-477 Waldron, Ingrid and Ricklefs, Robert E. *Environment and Population: Problems and Solutions*. New York: Holt, 1973. 232p.

A college textbook "concerned with human population growth and man's relationship to his environment." The book is intended as a text for an introductory biology course. References appear in the annotated bibliographies at the end of each chapter. Sections on food; resources, industrial production, and standard of living; natural processes; and human populations. Indexed.

5-478 Walters, A. Harry. *Ecology, Food & Civilisation*. London: Knight, 1973. 216p.

The author deals with world nutritional problems, urbanization, and the new industrial revolution, and predicts the survival of the human race. Using a historical approach, the author treats food supply as a determining factor in human relationships to the environment. A history, in effect, of the progress of civilization. Indexed. Bibliography.

5-479 Warner, Matt. *Your World—Your Survival*. New York: Abelard, 1971. 128p.

In a survey of the effects of a polluted environment and the waste of natural resources, the author covers overpopulation, water pollution, air pollution, land use, endangered species, noise pollution, and other problems, concluding with a call for humankind to make "peace with nature." Illustrated with black-and-white photographs. Brief bibliography. Indexed. Grades 5-8.

5-480 Warren, Charles E. *Biology and Water Pollution Control.* Philadelphia: Saunders, 1971. 434p.

A scholarly work written not only for biologists but also for other professionals, such as engineers, economists, and social scientists. It is a sophisticated but readable study, with a helpful introduction on the history of water pollution, standards, and the involvement of the biologist in this problem. Other major sections of the work cover the state of the aquatic environment, morphology and physiology, the ecology of the individual organism, and population and community ecology. A selected list of references accompanies each chapter. Illustrated. Indexed.

5-481 Watson, Richard A. and Watson, Patty Jo. *Man and Nature: An Anthropological Essay in Human Ecology.* New York: Harcourt, 1969. 172p.

According to the authors, this book "presents hypothetical models of eight ways of life—of the nonhuman primate, the protohuman, and six stages of man—synthesized from interaction of selected primates with the physical environment." The six stages of man used are elemental man, the advanced hunter and gatherer, the domesticator of plants and animals, the advanced food producer, industrial man, and atomic man. There is no index to this text on human ecology, but there is a lengthy bibliography.

5-482 Watt, Kenneth E. F. *Ecology & Resource Management: A Quantitative Approach.* New York: McGraw-Hill, 1968. 450p.

A technical book that deals with resource management from the standpoint of a mathematical or statistical method. Calculus and a first course in statistics are essential prerequisites for understanding many of the chapters, but the aim of the book is to reach an audience of biologists rather than mathematicians. The bulk of the book is devoted to developing analytic and simulation models for managing the environment. References accompany each chapter. The book is indexed by subject and author.

5-483 Watt, Kenneth E. F. *Principles of Environmental Science.* New York: McGraw-Hill, 1973. 319p.

A text on ecology written from the standpoint of environmental science in which the author treats not only plant and animal ecology, but a host of other disciplines as well. A number of applied sciences are treated at length, and elementary college-level mathematics has been used in presenting the material.

Illustrated. Chapters carry recommendations for further reading. Indexed.

5-484 Watt, Kenneth E. F. and others. *Systems Analysis in Ecology*. New York: Academic, 1966. 276p.

A book designed to survey the problems and techniques of systems analysis in ecology. Nine other authors have contributed to this book, with essays on small mammals, bird navigation experiments, and other topics, all technical in nature. The opening and closing chapters by the author are useful for defining this approach to ecology and the implications it has for science. Indexed.

5-485 Weisberg, Barry. *Beyond Repair: The Ecology of Capitalism*. Boston: Beacon, 1971. 256p.

The author relates ecological imbalance to social imbalance, seeing the two as part of the same problem. It is his position that neither the social nor the natural order can be restored within the present economic and political system. He gives particular attention to automobiles and petroleum, and the war machine. Indexed.

5-486 Wellford, Harrison. *Sowing the Wind: A Report from Ralph Nader's Center for Study of Responsive Law on Food Safety and the Chemical Harvest*. New York: Grossman, 1972. 384p.

A two-year study of the federal regulation of agribusiness that focuses primarily on the U.S. Department of Agriculture. Meat inspection, pesticide use and control, the functioning of House and Senate agriculture-related committees, and the influence of the chemical industry are exhaustively researched in this book. It carries an introduction by Ralph Nader and recommends the creation of a food safety agency and a consumer advocacy agency. Indexed.

5-487 Whiteside, Thomas. *The Withering Rain: America's Herbicidal Folly*. New York: Dutton, 1971. 224p.

This is the report of a serious journalistic investigation into the misuse of the herbicide 2, 4, 5-T and other chemicals in Vietnam and in the United States. The author documents government irresponsibility in failing to properly test and control defoliation agents later found to cause deforestation and problems of a medical nature. More than half the book consists of appendixes, which include government reports and documents about 2, 4, 5-T and other herbicides. Indexed.

5-488 Whittaker, Robert H. *Communities and Ecosystems*. New York: Macmillan, 1970. 162p.

This book offers background on the scientific concept of community, which is referred to in much of the literature about environmental problems. The first four chapters are concerned with "the structure of natural communities [and] the function of ecosystems," while the last two chapters deal with the problems of "man's relations to the biosphere." Each chapter includes a summary and a list of references. Illustrated. Indexed.

5-489 *Who Speaks for Earth?* By Barbara Ward and others. Edited by Maurice F. Strong. New York: Norton, 1973. 173p.

A series of seven lectures presented in 1972 that deal with the ocean, the economics of an improved environment, population, and environmental management, among other topics. Each lecture is briefly introduced. The lectures were sponsored by the International Institute for Environmental Affairs and were given in Stockholm at the United Nations Conference on the Human Environment, June 1972. Contributors featured are Barbara Ward, René Dubos, Thor Heyerdahl, Gunnar Myrdal, Carmen Miro, Lord Zuckerman, and Aurelio Peccei. Indexed.

5-490 Whyte, William H. *The Last Landscape*. Garden City, NY: Doubleday, 1968. 376p.

The author takes a positive approach to the problems of our major cities. He suggests that "desecration does seem a prerequisite for action" and contends that now people are ready to protect open space as a resource in cities. The book deals with present techniques for saving open space, current planning and development projects (and why they won't work), and the author's own prescription for ridding the cities of ugliness and overcrowding. Includes a bibliography. Indexed.

5-491 Widener, Don. *Timetable for Disaster*. Edited by Fay Robin Landau. Los Angeles: Nash, 1970. 277p.

The author, who wrote an award-winning television documentary on the environmental crisis in 1969, expands on the topic here in a popular account of major dangers and disasters: DDT, lead, smog, and Lake Erie's demise, among other things. Includes the transcript of the one-hour documentary, "The Slow Guillotine." This is a call to action. Brief list of suggested readings.

5-492 Wiens, Harold J. *Atoll Environment & Ecology*. New Haven: Yale University Press, 1962. 532p.

An exhaustive treatment of the ecology and environment of atolls by a geographer. This volume, in the author's words, is "largely limited to a dissection of the landscape and the physical and biological complex of the coral atoll in its tropical realm." The book contains many illustrations, a chapter on human beings in atoll ecology, a number of appendixes of items like average rainfall records, and an extensive bibliography. Numerous photographs end the book. Indexed.

5-493 Wilderness Conference, 6th, San Francisco, 1959. *The Meaning of Wilderness to Science*. Edited by David Brower. San Francisco: Sierra Club, 1960. 129p.

This Wilderness Conference was devoted to exploring the value of wilderness to the progress of science, and an international group of participants examined such topics as "plants and animals in natural communities," and "ecological islands as natural laboratories." Amply illustrated, but not indexed.

5-494 Wilderness Conference, 7th, San Francisco, 1961. *Wilderness: America's Living Heritage*. Edited by David Brower. San Francisco: Sierra Club, 1961. 204p.

A group of distinguished participants consider (1) wilderness and the molding of American character, (2) wilderness and the American arts, (3) the face of the land, and (4) wilderness resources. Illustrated, with biographical notes on contributors. Not indexed.

5-495 Wilderness Conference, 10th, San Francisco, 1967. *Wilderness and the Quality of Life*. Edited by Maxine E. McCloskey and James P. Gilligan. San Francisco: Sierra Club, 1969. 267p.

The proceedings of the 10th Wilderness Conference with papers such as "The Wilderness Act in Practice," "The Contribution of Wilderness to American Life," "Forgotten Wildernesses," and "The Quality of American Wilderness." Appendixes contain the text of the Wilderness Act of 1964, some federal regulations on wilderness management in the National Park Service, and biographical sketches of the conference participants. Not indexed.

5-496 Wilderness Conference, 11th, San Francisco, 1969. *Wilderness: The Edge of Knowledge.* Edited by Maxine E. McCloskey. San Francisco: Sierra Club, 1970. 303p.

These proceedings include the contributions of thirty-six participants. The conference focused on the "problems of the Alaskan wilderness and of wilderness and wildlife preservation everywhere." The editor has appended a list of endangered species, the text of the Endangered Species Conservation Act of 1969, Regulations on Public Outdoor Recreation Use of Bureau of Land Management Lands . . . , and a proposed National Conservation Bill of Rights. Biographical notes on contributors included. Not indexed.

5-497 Williams, Carrington Bonsor. *Patterns in the Balance of Nature and Related Problems in Quantitative Ecology.* New York: Academic, 1964. 324p.

A technical volume which deals with the problems of statistical ecology. There are essays on problems of species and small areas, problems of species and large areas, diversity as a measurable character of a population, and other topics. Mathematical. Indexed.

5-498 Winn, Ira J. *Basic Issues in Environment: Studies in Quiet Desperation.* Columbus, OH: Merrill, 1972. 488p.

A collection of readings intended for the college student and grouped around the following topics: man and nature; reflections on the land; life, liberty, and the pursuit of property; natural forces and human pressures; regional planning; the battle for wilderness; atomic challenge; the public interest and the New Conservation; the delicate balance; technology and the illusions of power; and the quality of life. Each section is accompanied by discussion questions and recommendations for further reading. Indexed.

5-499 Wollman, Nathaniel and Bonem, Gilbert W. *The Outlook for Water: Quality, Quantity, and National Growth.* Baltimore: Johns Hopkins University Press, 1971. 304p.

This book was published for Resources for the Future, and is an exhaustive study of water quality requirements and needs in the United States. Considerable tabular material is included in the technical explication of our water requirements. The study develops a "systematic economic model that (1) recognizes the regional aspects of the water problem yet yields a national perspective; (2) permits aggregation of demand and supply into usefully parallel concepts . . . and (4) identifies important choices to be made . . . among water resource regions." Indexed.

5-500 Wood, Nancy. *Clearcut: The Deforestation of America.* San Francisco: Sierra Club, 1971. 151p.

This is the story of a lost national heritage—the forests of the United States. The author is severely critical of the U.S. Forest Service and the timber industry. She documents the history of deforestation, past and current policies concerning national forests, the politics of the timber industry, and two major case studies in West Virginia and California. The book concludes with a call for a new forest policy.

5-501 Woodbury, Angus M. *Principles of General Ecology.* New York: McGraw-Hill, 1954. 503p.

Designed as an upper-level college text on general ecology, this book is divided into two major sections, on the physical environment and on biotic interrelationships. It is amply illustrated and contains an extensive bibliography. Indexed.

5-502 Woods, Barbara, ed. *Eco-solutions: A Casebook for the Environmental Crisis.* Cambridge, MA: Schenkman, 1972. 540p.

A book of readings centering around the following topics: population, air pollution, water pollution, noise pollution, land use, new towns, solid waste, esthetics, and environmental management. Notes on contributors. Not indexed.

5-503 Workshop on Global Ecological Problems, University of Wisconsin, 1971. *Man in the Living Environment; Report.* Madison: University of Wisconsin Press, 1972. 288p.

The proceedings of a Workshop on Global Ecological Problems held at the University of Wisconsin in 1971 by the Institute of Ecology (London) with support from the National Science Foundation. Four investigators organized the workshop: Robert F. Inger (The Institute of Ecology), Arthur D. Hasler (International Association of Ecology), F. Herbert Bormann (Ecological Society of America), and W. Frank Blair (U.S. International Biological

Program). A technical book, this report was sponsored so that an analysis of "important problems of environmental quality and management by a group of ecologists" could be transmitted to the 1972 United Nations Conference on the Human Environment. Recommendations approved by the entire workshop as well as individual "task group" recommendations appear in the back of the book.

5-504 Zwick, David and Benstock, Marcy. *Water Wasteland: Ralph Nader's Study Group Report on Water Pollution.* New York: Grossman, 1971. 494p.

A comprehensive, well-documented study of water pollution in the United States. The book analyzes the Water Quality Office of the Environmental Protection Agency, and is for the most part concerned with the role of the federal government in pollution control. However, the study also encompasses the "interrelationship of industry, citizen effort, and local government with the federal water pollution program." There are many appended documents, including a list of all Federal Water Pollution Enforcement Conferences from January 1957 through February 1971, composition of state pollution control boards, names of the members of the National Industrial Pollution Control Council, and approved water quality standards. Indexed, with extensive notes.

CHAPTER

⑥

GOVERNMENT PUBLICATIONS

————————————◆•◦•◆————————————

There is a wealth of information about the environment in government publications, though finding this information is frequently an arduous task. The first place to look is the *Monthly Catalog of United States Government Publications*, for information published by the federal government. Subjects to search include Air pollution, Ecology, Environment, Environmental Protection Agency, Pollution, Water pollution, and Water quality. In addition, other agencies in the federal government like the Council on Environmental Quality and the National Oceanic and Atmospheric Administration can produce substantial amounts of material. Some of the most useful information, however, will be found in House and Senate reports, and in hearings conducted by various House and Senate committees, such as the Senate Commerce Committee. While secondary sources such as the *National Journal*, *CIS Index*, and *CQ Weekly Report* will provide access to the subject content of a great deal of this material, the serious researcher will have to use the original sources, located through the *Monthly Catalog*.

In addition to U.S. government publications, other governmental publications have been included in this section because they merit inclusion in most environmental collections. Perhaps the most important of these is the report of the 1972 United Nations Conference on the Human Environment in Stockholm, Sweden. This conference generated worldwide excitement and was deemed by President Nixon to have been successful when the American proposal for an international ocean dumping convention and the establishment of a World Heritage Trust were approved. The report of this conference is contained in a United Nations document published in 1972.

6-1 United Nations Conference on the Human Environment, 1st, Stockholm, 1972. *Report of the United Nations Conference on the Human Environment Held at Stockholm, 5-16 June, 1972.* New York: Unipub, 1972. United Nations Document #A/Conf. 48/14. 122p.

The report of the Stockholm Conference on the Human Environment, including annexes.

While all the participating countries prepared reports on the status of the environment for the Stockholm Conference, two of these published documents particularly merit inclusion in environmental collections.

6-2 Great Britain. Department of the Environment. *Sinews for Survival: A Report on the Management of Natural Resources.* London: Her Majesty's Stationery Office, 1972. 74p.

A study of public opinion in the environment undertaken in connection with the United Nations Conference on the Human Environment. Sections on agriculture, forestry, fisheries, minerals, energy, water, wildlife, transport, and recreation. Recommendations are offered on each topic.

6-3 U.S. State Department. The Secretary of State's Advisory Committee for the 1972 U.N. Conference on the Human Environment. *Stockholm and Beyond.* Washington, DC: Government Printing Office, 1972. 152p.

The recommendation of a twenty-seven-member committee, chaired by Senator Howard Baker. Regional hearings were held and recommendations made on human settlements, resource management, pollutants, education, development, and institutional arrangements.

An earlier UNESCO document that will be of historical interest is contained in the proceedings of the following symposium.

6-4 Intergovernmental Conference of Experts on the Scientific Basis for Rational Use and Conservation of the Resources of the

Biosphere, Paris, 1968. *Use and Conservation of the Biosphere.* Paris: UNESCO, 1970. National Resources Research, No. 10. 272p.

Discussion in this conference was based upon reports from member states and ten review papers which had been prepared in advance for conference participants. These papers, revised in light of conference discussion, are included, along with the official report of the conference. The book also contains conference addresses of Malcolm S. Adiseshiah, M. G. Candau, A. H. Boerma, and Guy Gresford. A list of participants concludes the volume. Not indexed.

The following U.S. government publications provide useful information about ecology and will be of aid to librarians and researchers on the environment.

6-5 *Air Pollution Abstracts*, 1970–   . Monthly. Washington, DC: Government Printing Office. SuDoc #EP 4.11: (volume/number).

Prepared by the Office of Air Quality Planning and Standards of the U.S. Environmental Protection Agency, this bibliography catalogs the material received at the Air Pollution Technical Information Center (APTIC) in Research Tringle Park, North Carolina. Semiannual subject and other indexes are prepared, and abstracted articles are grouped into fourteen categories.

6-6 Council on Environmental Quality. *Environmental Quality*, 1970–   . Annual. Washington, DC: Government Printing Office. SuDoc #PrEx 14.1: (year).

A report to the Congress and to the nation, on the status of the quality of our environment, as mandated by Section 201 of the National Environmental Policy Act of 1969 (42 U.S.C. 4341). Information about progress, problems, and the whole range of environmental concerns. Illustrated. Indexed.

6-7 Council on Environmental Quality. *The Federal Environmental Monitoring Directory.* Washington, DC: Government Printing Office, 1973. SuDoc #PrEx 14.2: En 8/3. 105p.

A directory of federal environmental monitoring activities; also includes privately sponsored activity if it is especially complete or if there is no federal activity in the area. Sections on underlying factors, resources, ecological factors, pollution, manmade environment, and general sources. Names and telephone numbers are provided in this directory.

6-8 Council on Environmental Quality. *Ocean Dumping: A National Policy.* Washington, DC: Government Printing Office, 1970. SuDoc #PrEx 14.2: Oc2. 45p.

A report to the President on ocean dumping with findings and recommendations for a "comprehensive national policy on ocean dumping of wastes to ban unregulated ocean dumping of all material and strictly limit ocean disposal of any materials harmful to the marine environment."

6-9 Council on Environmental Quality. *The President's* [year] *Environmental Program,* 1970–   . Annual. Washington, DC: Government Printing Office. SuDoc #PrEx 14.2: En8/(year).

A collection which includes the President's State of the Union Message on Natural Resources and the Environment, proposed environmental legislation with letters of transmittal to the Congress, and analyses. In addition, a description is provided of initiatives which do not require legislation, or for which legislation is being proposed. Not indexed.

6-10 *Environmental Protection Research Catalog.* Prepared by the Smithsonian Science Information Exchange for U.S. Environmental Protection Agency, Office of Research and Monitoring, Research Information Division. Washington, DC: Government Printing Office, 1972. SuDoc #EP 1.2:R31/pts 1 and 2. 897p., 1,455p.

A catalog of research projects (see 3-e-4) based on research grants in government, industry, and universities. Research is grouped into the following categories: air quality, water quality, solid waste management, pesticides, radiation, and noise. Indexed by subject, investigator, performing organization, and supporting agency.

6-11 *How Can Our Physical Environment Best Be Controlled and Developed?* Compiled by the Legislative Reference Service, Library of Congress. Washington, DC: Government Printing Office, 1970. Senate Document 91-66. 231p.

Prepared for the national high-school debate topic for the school year 1970–1971. Under the question posed in the title, three debate propositions were advanced: *Resolved,* That the federal government should establish, finance, and administer programs to control air and water pollution in the United States; *Resolved*, That the federal government should establish economic penalties to control the pollution of our physical environment; and *Resolved*, That the federal government should control the use of all chemicals which pollute our physical environment. Reprints of material and the extensive bibliography relate to these three proposed topics.

6-12 National Science Foundation. National Science Board. *Environmental Science: Challenge for the Seventies*. Washington, DC: National Science Board, National Science Foundation, 1971. SuDoc #NS 1.28:71-1. 50p.

The third report of the National Science Foundation to Congress, this focuses on the "status and health of [environmental] science, as well as . . . the related matters of manpower. . . ." for 1971. Chapters on the future and the resources for environmental science. Its companion volume is *Patterns and Perspectives in Environmental Science* (6-13).

6-13 National Science Foundation. National Science Board. *Patterns and Perspectives in Environmental Science*. Washington, DC: National Science Board, National Science Foundation, 1972. SuDoc #NS 1.28: 73-2. 426p.

Prepared as a companion to 6-12, this provides the information and interpretation for the conclusions and recommendations reached in *Environmental Science*. Chapters on solar-terrestrial environment, dynamics of solid earth, climatic change, dynamics of the atmosphere-ocean system, severe storms, aquatic ecosystems, terrestrial ecosystems, environmental contaminants, human adaption and other topics. Illustrated and indexed. A bibliography of recommended readings is also included.

6-14 *102 Monitor*, 1971–   . Monthly. Washington, DC: Council on Environmental Quality. Available from Government Printing Office, SuDoc #PrEx 14.10: (volume/number).

Lists the environmental impact statements filed with the Council under the provisions of the National Environmental Policy Act as well as the Environmental Protection Agency's comment filed under the provisions of the Clean Air Act, items of concern in NEPA case law, and information on the Council's environmental studies.

6-15 *Selected References on Environmental Quality as It Relates to Health*, 1970–   . Monthly. Bethesda, MD: National Library of Medicine. Available from Government Printing Office. SuDoc #HE 20.3616: (volume/number).

A product of the MEDLARS (Medical Literature Analysis and Retrieval System) project, this bibliography cites selected articles on environmental quality as it relates to health in approximately 2,250 periodicals. Indexed by author and subject.

6-16 U.S. Environmental Protection Agency. *EPA Reports Bibliography*. Washington, DC: Government Printing Office, 1973. SuDoc #EP1.21 En8/5/973. 959p.

A listing of all EPA reports available from the National Technical Information Service as of April 1, 1973. Entries contain citation, abstract, corporate source, subject, contract, and author and title. A continuing bibliography.

6-17 U.S. Environmental Protection Agency. Division of Intergovernment Relations. Office of Legislation. *Environmental Program Administrators*. Washington, DC: Government Printing Office, 1974. SuDoc #EP 1.2:Ad 6. 42p.

A list, revised annually, of the directors of the following programs: state environmental, water quality, water supply, industrial waste, air quality, solid waste, noise, and pesticides. A list of interstate agencies having environmental program responsibilities is included as well as the EPA regional offices and regional administrators. Telephone numbers and addresses are included for all listings.

6-18 U.S. Environmental Protection Agency. Office of Research and Development. Washington Environmental Research Center. Environmental Studies Division. *Working Papers in Alternative Futures and Environmental Quality*. May, 1973. Washington, DC: Government Printing Office, 1973. SuDoc #EP 1.2:En 8/6. 242p.

Essays by a number of individuals on the nature of the environmental crisis, zero population growth, and the implications of alternative growth policies on environmental quality. Illustrated. Not indexed. This book was written for "policy-makers and planners" at all levels of government.

6-19 U.S. Environmental Protection Agency. Office of Research and Development. Washington Environmental Research Center.

Environmental Studies Division. *Final Conference Report for the National Conference on Managing the Environment.* Washington, DC: Government Printing Office, 1973. SuDoc #EP 1.2:M31. 274p.

The proceedings of a conference focusing on the environment as a public policy issue, held May 14–15, 1973, in Washington, DC.

An additional bibliography produced by the government and available from the National Technical Information Service will be useful to researchers on problems of the environment. This bibliography monitors the research formally presented in *Government Reports*.

6-20 *Environmental Pollution and Control*, 1969–   . Weekly. Springfield, VA: National Technical Information Service, U. S. Dept. of Commerce. SuDoc #C51.9/5: (number).

CHAPTER

NONPRINT MEDIA

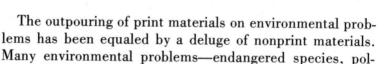

The outpouring of print materials on environmental problems has been equaled by a deluge of nonprint materials. Many environmental problems—endangered species, polluted rivers and lakes, littered parks, clearcut forests—are most poignantly and effectively presented through nonprint media.

The compilation on nonprint materials presented here is extremely selective. We have included only those materials considered outstanding in form and content, and of general interest to a broad range of users. Omitted from consideration were materials covering very narrow subject matter, such as an individual endangered species. Our selected listing includes 16mm films, filmstrips, recordings, simulation games, slides, and kits. Eight-millimeter films are not included.

Nonprint materials are often quite expensive, and therefore we assume that librarians will want to purchase these materials only after careful consideration. We have included prices of nonprint materials as an additional aid to comparative evaluation. The materials in the following listing have all received excellent reviews. In the case of 16mm films, we offer two levels of rating, "highly recommended" and "recommended." The notation "general" in the bibliographic statement indicates that the material can be used effectively with people fourteen years old and up in either instructional or noninstructional settings.

156

16MM FILMS

Highly Recommended

7-a-1 *Before the Mountain Was Moved.* 58 min., color, 1971. $595 purchase. Distributed by Contemporary/McGraw-Hill Films. General audience.

Produced originally for the U.S. Office of Economic Opportunity, this film was nominated for an Academy Award as best documentary film of the year. It is the story of West Virginia coal miners and farmers who, having suffered avalanches and floods, fought the strip mining companies and eventually succeeded in getting legislation to prevent indiscriminate strip mining. The people play themselves in this movie, which documents the power of a real grass-roots movement.

7-a-2 *A Fire.* 18 min., color, 1969. $145 purchase, $9 rent. Produced by Ephraim Golestan. Distributed by Texture Films, Inc. General audience.

This film deals with the relationship between developed and underdeveloped nations and the effects of an ecological disaster on human lives. An oil well in the Iranian desert is juxtaposed with the subsistence farming carried on in the derrick's shadow. The well in no way improves the lot of the farmers and, in fact, endangers their lives when it erupts and catches fire. The movie was filmed by an all-Iranian crew.

7-a-3 *For Your Pleasure.* 4 min., color, 1971. $100 purchase, $10 rent. Distributed by Mass Media Associates. General audience.

A highly imaginative wordless cartoon, this brief film uses John Constable's painting *The Hay Wain* as its setting. Dozens of animated characters stroll through the landscape. Then the little people produce tiny buzz saws and cut down painted trees, boats begin to fill the river, and a city begins to grow on the horizon. Finally chaos erupts, and the city overflows the painting and spills out into the museum. Effective presentation for both children and adults.

7-a-4 *Home.* 29 min., color, 1972. $300 purchase. Produced and distributed by the Southern Baptist Convention, Radio and TV Committee. General audience.

This film compares a speech given in 1855 by Chief Seattle of the Duwanish Tribe and the current American environment. Responding to a federal request that his tribe cede their lands to territorial authorities, Chief Seattle argued against desecration of the land, concluding that the white man's appetite "will devour the earth and leave behind only a desert." While the narrator is reading Seattle's eloquent plea, the camera records examples of environmental destruction caused by our cultural obsession with economic progress.

7-a-5 *No Room for Wilderness?* 26 min., color, 1968. $315 purchase, free loan. Sponsored by the Sierra Club. Distributed by Associated-Sterling Films. General audience.

This film documents the destruction of a South African wilderness that disappeared as the area became "civilized." Beginning with scenes of the wilderness as it was originally, the film chronicles colonization, which brought sheep and cattle into the ecosystem, upsetting the balance of nature. Streams dry up, species disappear, and the wilderness continues to decline. The final message is that population control is the only way to preserve both wilderness and mankind.

7-a-6 *Of Broccoli and Pelicans and Celery and Seals.* 30 min., color, 1970. $315 purchase, $11.50 rent. Sponsored by the Ford Foundation and Old Dominion Foundation. Distributed by Indiana University Audio Visual Center. General audience.

This film presents a critical and factual analysis of the effects of DDT on the elephant seals, pelicans, and other creatures living on the Channel Islands, protected refuges just off the California coast. Scientists noted a drastic change in the seals' behavior when the mother seals began killing their pups. The aberration was traced to DDT used by farmers in nearby broccoli and celery fields. No easy solutions are offered, but provocative questions are raised. Useful from the junior-high level on up.

7-a-7 *Pollution Is a Matter of Choice.* 53 min., color, 1970 (released for distribution in 1973). $500 purchase, $25 rent. Produced and distributed by NBC Educational Enterprises. General audience.

In this award-winning documentary narrated by Frank McGee, the ecological dilemmas of two communities are explored. The people of Machiasport, Maine, must decide whether to allow their seaport to become a dock for supertankers. The people of the area badly need the jobs and money this would provide, but a heavy environmental price would have to be paid. A choice must be made. The second choice outlined in the film is whether or not to build an airport in the vicinity of the Everglades in Florida. This film was a winner at the American Film Festival and won the Cine Golden Eagle.

7-a-8 *The Redwoods.* 20 min., color, 1968. $250 purchase, free loan. Sponsored by the Sierra Club. Distributed by Associated-Sterling Films. General audience.

This simple but effective film documents the demise of the redwoods which once covered the entire northern half of the earth's surface and now are in danger of extinction. Moving from a long, slow look at a magnificent redwood to the devastation and emptiness of a logged area, the film is an eloquent statement of loss.

7-a-9 *The Rise and Fall of the Great Lakes.* 17 min., color, 1968. $200 purchase, $15 rent to U.S. users. Produced and distributed by the National Film Board of Canada. General audience.

In an amusing chronicle of the history of the Great Lakes, a canoeist travels back through time, finding himself stuck on a glacier. The film portrays the historical, geological, and ecological life of the lakes, juxtaposing past beauty with present pollution. Folk ballads and the humorous tone of the film in no way lessen the strength of the ecological point.

7-a-10 *Tomorrow Is Maybe.* 60 min., color, 1972. $550 purchase, $20.50 rent. Distributed by Indiana University Audio Visual Center. General audience.

A thoughtful, rather pessimistic analysis of the human relationship to the environment, this film focuses on overpopulation as our most critical immediate problem. In a narrative sprinkled with literary and scholarly quotations, the film points out that technological advancement has outdistanced moral growth, but that the answer is not a return to the past. Using the analogy of rats leaving a sinking ship, the film notes that people, unlike rats, have the ability to understand and therefore affect their own future. Whether or not they will is left an open question.

7-a-11 *Tomorrow's Children.* 17 min., color, 1972. $225 purchase, $23 rent. Distributed by Perennial Education. General audience.

According to this film, "tomorrow's children" will never know an uncrowded world. The film provides an overview of the way population growth has interfered with the ecological balance. The narrator discusses life cycles and the harmony of nature and points out the high consumption level in the United States compared to other countries. Finally, the film offers some alternative solutions to the problems it has outlined.

7-a-12 *Tragedy of the Commons.* 23 min., color, 1971. $295 purchase. Distributed by Holt. General audience.

This controversial, provocative film begins with the essay, "The Tragedy of the Commons," by Garrett Hardin, and goes on to explore the effects of overpopulation on people, the inequity in the distribution of resources in the United States and the world, and possible solutions to the problem. The first part of the film is a drama, set in eighteenth century England, which demonstrates how resources such as common pastureland are quickly depleted when they are used as if inexhaustible. In the second part of the film, Hardin discusses the effects of overcrowding; advertisements for American consumer products are followed by shots of dumps, slums, etc. The third part of the film deals with overcrowding and stress. The final sequence may be offensive to some audiences because it suggests that maternity is not a virtue. Hardin concludes that laws to restrict population growth are necessary, but he does not specify what laws should be enacted. A study guide for the film is available.

Recommended

7-a-13 *Energy: A Matter of Choices.* 22 min., color, 1973. $265 purchase. Distributed by Encyclopedia Britannica Educational Corporation. Grade 5 through general audience.

This film presents the difficult choices to be made between energy needs and environmental protection, offering no easy solutions to the problem. The film documents the history of energy consumption in the United States and points out how contradictory and inconsistent have been simultaneous demands to meet energy needs and save the environment. The true costs of energy are explained, including environmental costs, and an estimate is made of when the present sources of U.S. energy will run out. Should provide basis for good discussion in the classroom or a public program.

7-a-14 *Epilogue.* 16 min., color, 1972 (U.S. release). $235 purchase, $25 rent to U.S. users. Produced by the National Film Board of Canada. Distributed by Benchmark Films. General audience.

Technology is contrasted with nature's ecological balance.

Beautiful scenes of lakes, mountains, and animals are contrasted with man's contributions—buildings, automobiles, billboards, traffic jams, pollution, and destruction of wildlife. The story of man's overwhelming, destructive touch is told in song. The film ends with a poem written in 1856 by Henry David Thoreau.

7-a-15 *Keepers of Wildlife.* 21 min., color, 1974. $275 purchase for U.S. buyers, apply for rental. Produced by National Film Board of Canada for Canadian Wildlife Service. Distributed by ACI Films. General audience.

This is a positive record of Canadian Wildlife Service efforts to save endangered species. A good balance to the more hysterical films which document doomed animals and leave little hope for improvement, this film shows what concerned people have done to restore the environment.

7-a-16 *Let No Man Regret.* 11 min., color, 1973. $140 purchase. Produced and distributed by Alfred Higgins Productions. Elementary through general audience.

Without narration, this film speaks clearly about man's destruction of natural beauty. The camera slowly surveys a beautiful, quiet wilderness area—the birds, flowers, waterfalls, butterflies—and then the blast of a single horn introduces civilization. Destruction ensues, with initials carved into trees, bullet holes everywhere, and a devastating forest fire.

7-a-17 *Nails.* 4 min., color, 1973. $75 purchase, $8 rent. Distributed by International Film Bureau. Elementary through general audience.

This film comments on the dangers of overpopulation, using animation without narration. All sorts of nails begin to fill the screen. New clusters of nails form until the space is entirely filled. Then the nails begin to rust and disintegrate. Winner of the Cinegram Prize for Swiss animation.

7-a-18 *One Hand Clapping.* 9 min., color, 1973. $120 purchase, $10 rent to U.S. users. Produced by the National Film Board of Canada. Distributed by Macmillan Films. General audience.

Introducing the phenomenon of sound pollution, this film shows deaf children in therapy and older people learning of their deafness. It goes on to present noise in the abstract and its ill effects, using the overwhelming, painful sounds of jackhammers, car horns, and other equipment to illustrate noise pollution. Makes good use of its nine minutes.

7-a-19 *Population and the American Future.* Two 30 min. segments, color, 1972. $300 purchase. Distributed by Population Commission Films. General audience.

This filmed version of a research report presents the Commission on Population Growth and the American Future's recommendations and findings. The commission was established by Congress and appointed by the President for a two-year investigation. The film, narrated by Hugh Downs, is low-key and direct. Sex education, abortion, and contraception are considered, as well as environment, city life, and education. A teacher's guide is available.

7-a-20 *The Ravaged Earth.* 27 min., color, released 1970. $300 purchase, $15 rent. Produced by WKYC, NBC Cleveland. Distributed by NBC Educational Enterprises. General audience.

This film shows how strip-mining processes have ravaged the earth. The camera is used effectively to show devastation in Ohio and other states, the barren land scarred with slag heaps and silted streams. Stewart Udall comments on the need for regulation and reclamation. Winner at the American Film Festival.

7-a-21 *Recycled Reflections.* 11½ min., color, 1973. $145 purchase, $15 rent. Distributed by Film Fair Communications. Grade 5 through general audience.

This film is shot at a car bumper recycling plant. Without narration, the film moves through the piles of wrecked cars to the plant where workers are welding, sanding, cleaning, etc. An artful, positive look at one solution to our refuse problem.

7-a-22 *Urban Impact on Weather and Climate.* 16 min., color, 1972. $225 purchase, $20 rent. (Series: Environmental Sciences Series.) Distributed by Learning Corporation of America. General audience.

Illustrated with views of urban areas, time-lapse sequences of laboratory experiments, drawings, and wind charts, this film reports how human beings have altered the environment, as recorded by the American Meteorological Society. Higher summer temperatures, lower winter temperatures, more rainfall, flooding, fog and dangerous recirculation of polluted air are the immediate results of congestion, large buildings, highways, and loss of greenery. Long-range problems are also discussed, with suggestions for countering or reducing the negative influences of civilization on the weather and climate.

FILMSTRIPS

7-b-1 *Checkmate: Our Use and Misuse of the Land.* 2 filmstrips, each with phonodisc or cassette, with teacher's guide, color, 1974. $37.50 with discs, $41.50 with cassettes. Distributed by Audio Visual Narrative Arts. Grade 7 through general audience.

The first filmstrip in this set discusses the delicate balances in nature, human interruption of that balance through population growth and industrial growth, and how the American economic system has promoted destruction of the land. The second strip analyzes the problem of growth more carefully, with discussions of automobiles, highways, waste disposal and population concentration. The narrator points out how the tax structure and government policies have strengthened the industries which are major polluters. The strip ends on a positive note, suggesting that there are solutions to be found.

7-b-2 *Crisis of the Environment.* 5 filmstrips with phonodiscs, color, 1970. $97.50 series. Distributed by New York Times, Educational Division. Grade 7 through general audience.

Five aspects of the environmental crisis are treated here, with emphasis placed on the role of values in creating and solving the problems. The five filmstrips discuss man as an endangered species, the dilemma caused by the use of pesticides to increase food production, the clash of viewpoints about saving wildlife, the economic benefits of progress vs the value of conserving resources and the dimensions of the population problem.

7-b-3 *The Ecological Crisis.* 6 filmstrips, with 3 phonodiscs or cassettes, color, 1971, $69 with discs, $72 with cassettes. Distributed by QED Productions. Grade 7 through general audience.

The filmstrips deal with population statistics and trends, ecological considerations, evolution and extinction, pesticides, and pollution. The material is well organized for either classroom study or citizen group use.

7-b-4 *Ecology: Interrelationships in Nature.* 8 filmstrips, with 4 phonodiscs or cassettes, color, 1974. $84 with discs, $91 with cassettes. Distributed by Society for Visual Education. Grade 6 through general audience.

This series is biological rather than sociological in approach, focusing in each strip on a different type of natural environment, such as swamp, desert, grassland, or forest. The filmstrips show how sun, water and land·affect plant life, which in turn determines animal populations. The last filmstrip in the series, "Man-made Communities," discusses how human beings interrupt natural cycles and the problems which such interruption brings.

7-b-5 *Environmental Studies Series.* 6 filmstrips, each with phonodisc or cassette, with teacher's guide, color, 1972. $72.50 with discs, $84.50 with cassettes. Distributed by Centron Educational Films. Grade 6 through general audience.

The series deals with air, water, and noise pollution, solid waste disposal, land management, and the use of pesticides. Causes of the various forms of pollution are considered, as well as ways of controlling them through research, legal restrictions, government controls, and civic and individual action.

7-b-6 *Exploring Ecology.* 5 filmstrips, each with cassette or phonodisc and guide, color. 1974. $67.50 series. Distributed by National Geographic Society. Grade 7 through general audience.

These filmstrips explore five ecosystems: mountains, rivers, woodlands, prairies, and swamps. Through National Geographic's always excellent photography, the viewer is offered a poetic introduction to each of the environments in its natural state, details of the plant and animal life common to that environment, and documentation of environmental destruction endangering that ecosystem. Each filmstrip is self-contained and could be used very effectively by itself.

7-b-7 *Man: An Ecological Approach.* 18 filmstrips, with 10 cassettes and 5 guides, color, 1972–73. Released 1974. $206 series, units also separately priced. Distributed by Educational Resources Division, Educational Design. Grade 7 through general audience.

A well-organized, interdisciplinary approach to humanity's ecological problems, this filmstrip series places our environmental crisis in an international perspective. While individual units, such as "Eco-problems" and "Eco-action," could be used separately, the strength of the filmstrips is in the integration of the series. After an introductory unit, "Earth," there are supporting units entitled "Topography," "Vegetation," "Basic Resources," "Technological Resources," and "Climate" (the last is particularly good).

7-b-8 *New York City: An Environmental Case Study.* 2 filmstrips, each with phonodisc and teacher's guide, color, 1971. Released 1972. $34. Produced and distributed by Denoyer-Geppert Audio Visuals. General audience.

A thoughtful presentation of the environmental crisis in New York City, exploring the results of population growth and concentration, industrial expansion, waste, and mass transportation, as well as the few remaining green areas and a wildfowl refuge in the city. The filmstrips point out that environmental destruction goes beyond air and water pollution to destruction of the human spirit. People lose the capacity even to see the ugliness of their surroundings.

7-b-9 *Spaceship Earth.* 6 filmstrips, each with cassette and teacher's guide, color, 1972 (released 1973). $120 series, $20 each. Distributed by Hawkhill Associates. Grade 5 through general audience.

Producer and narrator Bill Stonebarger pits his own naturalist philosophy against what he sees as an overly specialized technological society. Folk music background does not detract from the scientific information provided in five strips on the universe, the biosphere, living things, the cell, atoms and molecules, and a final "think piece" called "A Little While Aware."

7-b-10 *Squandered Resources.* Filmstrip, with phonodisc and teacher's guide, black and white, 1971. $9. Distributed by New York Times, Educational Division. Grade 8 through general audience.

This filmstrip presents a historical summary of the conservation movement and the use and abuse of resources in the United States. The survey of environmental problems is illustrated with news photos, graphs, and cartoons. The filmstrip warns against continued misuse of the environment on which we depend for survival. The reverse side of the disc is a conversation between a scientist and students.

7-b-11 *Surviving the Ecology Crisis*. 4 filmstrips, with 2 phonodiscs or cassettes and teacher's guide, color, 1972. $36 with discs, $40 with cassettes. Distributed by Society for Visual Education. Grade 6 through general audience.

This series provides the viewer with an international perspective on the sources and effects of environmental destruction, overpopulation, poverty, and the diminishing availability of power resources. The filmstrips deal with both personal and societal consequences of pollution and the abuse of resources.

7-b-12 *Waterways or Sewers*. 4 filmstrips, with 4 phonodiscs or cassettes and teacher's guide, color, 1972. $60 series. Distributed by Time/Life Education. General audience.

This comprehensive presentation on the seriousness of water pollution, especially of the Great Lakes, offers some solutions but emphasizes that economics is the issue. With industrialization comes pollution; difficult choices must be made if natural resources are to be saved. Pollution of the Hudson River is also documented in this series.

RECORDINGS

7-c-1 *Earth Day*. A celebration and lament for voice and orchestra read by Stacy Keach with music from Ralph Vaughan Williams. Phonodisc, 62½ min., $6.98; cassette $7.95. 1974. Distributed by Caedmon Records. General audience.

Using words and music to complement each other, this is a celebration and lament for the earth. With selections from Whitman, Thoreau, Shakespeare, Eliot, Rachel Carson, Dickinson, Darwin, and John Muir, the recording is an anthology on the earth—its problems and its promise.

7-c-2 *Ecological Spectrum*. Cassettes, each 25-50 min., each $14.95. 1972. Distributed by Center for Cassette Studies. General audience.

This is a collection of over twenty-five cassettes covering a broad range of environmental problems. Some of the titles are: "Noise Pollution," "Insecticide Treadmill," "Troubled Waters," "Ecology and the GNP," and "Man: Friend or Foe?"

7-c-3 Ehrlich, Paul R. *Population Growth and the Ecological Balance*. Audiotape reel-to-reel, 3¾ ips, 60 min., $10; cassette, $10.

1970. Produced and distributed by Big Sur Recordings. General audience.

The well-known author of *The Population Bomb* (5-159) here repeats the speech he gave at the First World Hunger Conference in 1969. In it he outlines the serious conditions of pollution and overpopulation in the world. Ehrlich is critical of the U.S. Department of Agriculture and the Congress for failing to deal effectively with these problems. The forty-minute speech is followed by a twenty-minute question period, in which Ehrlich responds to questions about overpopulation, especially in China. Requires careful listening, but Ehrlich's wit provides the needed breaks.

7-c-4 *Profit and Pollution.* Cassette, 59 min., $16.95. 1972. Produced and distributed by Center for Cassette Studies. General audience.

This a recorded lecture by Barry Commoner in which he discusses changes in technology that have worked to the detriment of the environment. According to Commoner, technology has provided us with more than we need and, in the process, has disrupted our relationship with nature. Study questions and suggested readings are included.

SIMULATION GAMES

7-d-1 The Planet Management Game. Playing time: 3-4 hours. Number of players: 2-10 in groups. $17. Distributed by Houghton Mifflin. Grades 7+.

In this game, the players govern the planet Clarion and attempt to improve living conditions there. Players learn how one action affects another while dealing with a wide range of ecological problems. Solutions become complex in the game, as they are in reality.

7-d-2 The Pollution Game. Playing time: 1-2 hours. Number of players: 5. $13. Distributed by Houghton Mifflin. Grades 5-7.

This game teaches the economic and political dimensions of environmental problems. Each player owns industries on a game board. The industries pollute the environment, but the players are also responsible for protecting the environment. Players are supposed to earn as much money as possible, and at the same time prevent the water and air pollution indexes from rising too high. Everyone loses if the indexes reach lethal limits.

7-d-3 Population: A Game of Man and Society. Playing time: 1-2 hours. Number of players: 2-6. $10. Distributed by Damon Corporation. Grades 7+.

Players learn basic concepts of resource utilization and population control. They must lead an imaginary country through three stages of development, using a plan of controlled growth so that population does not exceed productivity.

SLIDES AND KITS

7-e-1 Environmental Problems Resources Kit. Cardboard mounts, color, 40 slides, and teacher's guide. 1973. $39. Distributed by Educational Resources Division, Educational Design. Grades 5+.

This kit includes twenty pairs of slides, one diagrammatic and one photographic, for each topic. The pairs of slides can be used in any order. Each page of the guide corresponds to one set of slides, giving background information on each topic and presenting theories to be discussed.

7-e-2 Environmental Pollution. Carboard mounts, color, 2 units with 20 slides each, and teacher's guide. 1973. $9 each unit. Distributed by Society for Visual Education. Grades 3-7.

The two units included here are "Environmental Pollution" and "Air and Water Pollution." The slides represent introductory material and the guide offers a text to be read with them.

7-e-3 Johnny Horizon Environmental Test Kit. $10. Distributed by Parker Brothers.

In 1973, Consumers Union tested fourteen environmental demonstration and testing kits. This is the only one they recommended, and then with the caveat that it contains one experiment on culturing microorganisms which they feel should be done only under professional supervision. The kit contains nine other experiments on physical and chemical substances in air and water. Supplies and chemicals were considered well labeled, and the instructions, clearly stated.

CHAPTER

PERIODICALS

The environmental periodical, for the most part, has proved to be short-lived. Excellent magazines such as *Clear Creek* and *Environmental Quality* ceased publication early in the seventies for a variety of reasons; since that time, no new magazines have been founded that focus exclusively on the environment and are intended for the general reader. However, a number of conservation magazines have begun to devote more attention to environmental issues, and so have popular periodicals like *Harper's Magazine* and *Saturday Review*. What remains today, then, in the way of regular coverage, is a substantial core of conservation magazines offering more and more information about the environment.

In preparing this periodicals list, we have sought for publications which present the nontechnical aspects of environmental problems. For this reason *Ecology Law Quarterly* and *Environmental Law Review* have not been included. Although the *Bulletin of the Atomic Scientists* can be found in a number of libraries, it is not listed here because of its focus. A number of technical periodicals were carefully studied but omitted for similar reasons.

8-1 *Ambio: A Journal of the Human Environment*, 1972–
Bimonthly. Boston: Universitetsforlaget.

Published in English, this journal provides popularly written material and technical information in its "Reports" section.

8-2 *American Forests*, 1895–   . Monthly. Washington, DC: American Forestry Association.

Monthly features include a "Washington Lookout" column and

an editorial. Most articles pertain to forestry, though many are considerably broader in scope. The American Forestry Association is an independent and apolitical national conservation association.

8-3 *Audubon Magazine*, 1899–　　. Bimonthly. New York: National Audubon Society.

A periodical that focuses on the conservation of all types of natural species, not only birds.

8-4 *Catalyst for Environmental Quality*, 1871–　　. Quarterly. New York: Catalyst for Environmental Quality.

This small magazine contains informative essays on issues of pressing environmental concern.

8-5 *Conservation Foundation Letter*, 1966–　　. Monthly. Washington, DC: The Conservation Foundation.

Most issues are devoted to a single topic of environmental concern. Particularly useful for keeping abreast of federal environmental control actions.

8-6 *Defenders of Wildlife News*, 1926–　　. Bimonthly. Washington, DC: Defenders of Wildlife.

A magazine with a number of special features, including a section on predator control, endangered wildlife, and wildlife bills in Congress. It also carries regular columns with feature essays by many well-known conservation writers.

8-7 *Ecologist*, 1970–　　. Ten issues per year. Wadebridge, England: Ecologist.

A leading British magazine that covers environmental issues in all parts of the world, in addition to current developments in the United Kingdom.

8-8 *Environment*, 1958–　　. Monthly. St. Louis: Scientists' Institute for Public Information.

Environment features reports on the environment from all over the world. Previously entitled *Scientist and Citizen*, it presents authoritative articles on every issue that relates to the environment.

8-9 *Environmental Action,* 1969–   . Biweekly. Washington, DC: Environmental Action, Inc.

Provides news of federal environmental concerns, as well as short informative articles about numerous environmental issues.

8-10 *Environmental Affairs,* 1971–   . Quarterly. Brighton, MA: The Environmental Law Center.

A quarterly devoted to scholarly articles that cover a considerable range of environmental problems. Articles on international aspects of environmental pollution as well as case studies of individual cities in the United States.

8-11 *Environmental Science and Technology,* 1967–   . Monthly. Washington, DC: American Chemical Society.

The chemical approach to environmental technology with interviews, feature articles, summaries of research, and news. The *Pollution Control Directory,* issued each year in October, is part of this magazine.

8-12 *International Wildlife,* 1971–   . Bimonthly. Washington, DC: National Wildlife Federation, Inc.

A magazine devoted to covering, in a popular fashion, wild animals in all parts of the world and the problems that are connected with their survival. A companion magazine to *National Wildlife* (8-17).

8-13 *Journal of the Air Pollution Control Association,* 1951–   . Monthly. Pittsburgh: Air Pollution Control Association.

A technical periodical that covers all aspects of air pollution. Each issue contains a number of technical papers, in addition to general articles such as "A Review of the Background, Preparation and Use of Environmental Impact Statements."

8-14 *Journal of Environmental Education,* 1970–   . Quarterly. Washington, DC: Heldref Publications.

A publication devoted to education about the environment. The editor is a professor of journalism and wildlife ecology at the University of Wisconsin—Madison. Useful for locating techniques and examples employed in conservation instruction.

8-15 *Living Wilderness*, 1935–   . Quarterly. Washington, DC: The Wilderness Society.

Articles about wilderness and the animal life that depends upon wilderness. Reports agency and congressional action on environmental issues.

8-16 *National Parks and Conservation Magazine*, 1919–   . Monthly. Washington, DC: National Parks Association.

Published by the National Parks Association, an independent public-service organization that attempts to protect and restore the national parks and monuments of America. Contains extensive material on our national park system.

8-17 *National Wildlife*, 1962–   . Bimonthly. Washington, DC: National Wildlife Federation, Inc.

A companion magazine to *International Wildlife* (8-12), this magazine focuses on the environment and our native animal population. Features include reports of legislation at the national and state levels.

8-18 *Ranger Rick's Nature Magazine*, 1967–   . Monthly. Washington, DC: National Wildlife Federation, Inc.

A children's magazine that deals with environmental and conservation concerns.

8-19 *Sierra Club Bulletin*, 1893–   . Monthly. San Francisco: Sierra Club.

The official publication of the Sierra Club features news of the organization, a number of general essays, and roundups of legislative and environmental information.

CHAPTER

ORGANIZATIONS

Among the best current and continuing sources of information about environmental problems are the organizations devoted to the environment and its preservation. These include private environmental organizations, libraries, research organizations, and environmental consultants.

The organizations included in this chapter have all responded to a questionnaire, an outline of which is included with each type of organization. The information provided about each group is an edited version of that group's response to the questionnaire.

Private Organizations

Approximately 15 percent of the private, nonprofit organizations we contacted are no longer in existence, perhaps an indication that the glow of Earth Day has worn off for some enthusiasts or perhaps simply a sign of the economic squeeze. In any case, the thirty-eight organizations listed here are very much alive and well. As our questionnaire summary indicates, we asked each group if it provides reference or other services to libraries. Thirty-three of the thirty-eight groups indicated that they do answer reference questions. We have included the groups that responded negatively in order to spare both librarians and those organizations wasted effort. (One ground rule on reference questions by telephone: It is assumed that the inquiring library will bear the cost of the call.)

Private organizations are important sources of pamphlets and monographs. While we could not list the priced publications available from each organization, most of these groups have publications lists which are available upon request. Many of their publications cost less than $2.

The questionnaire sent to private organizations requested the following information: name, address, telephone number, date established, priorities or focus, publications, requirements for and advantages of institutional membership, services rendered to libraries, and additional information.

9-a-1 Air Pollution Control Association
4400 Fifth Avenue
Pittsburgh, PA 15213
Telephone: 412-621-1090

Established: 1907

Priorities: Devoted solely to air pollution control.

Institutional membership: Open to libraries. Organizations performing technical, scientific, research, education, consultant, or civic services interested in accomplishing the purposes of the association are eligible as Organization Members.

Serial publications: *Journal of the Air Pollution Control Association*, monthly; *Directory of Governmental Air Pollution Agencies*, annual; *Proceedings Digest*, annual.

Services to libraries: Will answer technical reference questions on air pollution control by telephone or mail. Contact Mr. Edward F. Frederick.

9-a-2 American Forestry Association
1319 18th Street NW
Washington, DC 20036
Telephone: 202-467-5810

Established: 1875

Priorities: Advancement of intelligent management and use of our forests, soil, water, wildlife, and all other natural resources.

Institutional membership: Open to libraries. $7.50 per year. Members receive monthly magazine, and books at discount.

Serial publications: *American Forests*, monthly.

Services to libraries: Will answer reference questions by telephone or mail. Contact Mr. Pardo. Vertical file material on forestry and conservation free upon request.

9-a-3 American Littoral Society
Sandy Hook
Highlands, NJ 07732
Telephone: 201-872-0200

Established: 1961

Priorities: Coastal zone environment.

Institutional membership: Open to libraries. Members receive all publications, newsletters.

Serial publications: *Underwater Naturalist*, quarterly.

Services to libraries: Will answer reference questions by phone or mail. Contact Executive Director. Newsletters and booklists are free upon request.

9-a-4 Center for Study of Responsive Law
PO Box 19367
Washington, DC 20036
Telephone: 202-833-3400

Established: 1968

Priorities: Bringing to public and official discussion issues such as air and water pollution and food and pesticide regulation. Current projects include a major conference on nuclear power and The Clean Water Action Project.

Institutional membership: Not a membership organization.

Services to libraries: Will answer reference questions by telephone or mail.

Additional information: This group is the research and investigative arm of Ralph Nader and his citizen advocates. The Center also serves as an information resource, supplying data and help to the press, government officials, and concerned individuals.

9-a-5 Committee for Environmental Information
438 North Skinker Boulevard
St. Louis, MO 63130
Telephone: 314-863-6560

Priorities: Dissemination of scientific information on the environment.

Institutional membership: Open to libraries. Members may use organization's library and attend all meetings and workshops.

Serial publications: *C. E. Eye Newsletter.*

Services to libraries: Will answer reference questions by telephone (if not too involved) and by mail. Contact Sarah Miller, Director Information Services. *C. E. Eye Newsletter* available free upon request.

Additional information: "At present our Scientific Division . . . of forty scientists in the St. Louis area are working on the energy problem, offering expertise on the issue of nuclear power in Missouri. We hope to be funded to provide information on the lead problem in the St. Louis area. Other projects are the preparation of information on the shipment of hazardous materials and the effects of photochemicals on the ozone layer."

9-a-6 The Conservation Foundation
1717 Massachusetts Avenue NW
Washington, DC 20036
Telephone: 202-265-8882

Established: 1948

Priorities: Land-use policy studies.

Institutional membership: Not a membership organization.

Serial publications: *Conservation Foundation Letter*, monthly.

Services to libraries: Will answer reference questions only by mail. Contact Patricia Hartman, Information Assistant. Some vertical file material available free upon request.

9-a-7 Ecology Center, Inc.
2179 Allston Way
Berkeley, CA 94704
Telephone: 415-548-2220

Established: 1969

Priorities: An environmental information clearinghouse, with library and switchboard services. Also developing alternative programs.

Institutional membership: Open to libraries. $12 per year.

Serial publications: *Ecology Center Newsletter*; *Bay Area Environmental Calendar.*

Services to libraries: Will answer reference questions by telephone or mail (send postage please). Contact Rick Codina. Fact sheets available for cost of postage.

9-a-8 Ecology Center of Southern California
2315 Westwood Boulevard
Los Angeles, CA 90064
Telephone: 213-475-1619

Established: 1972

Priorities: Environmental clearinghouse providing information, reference, and referral services on ecological issues.

Institutional membership: Open to libraries. $10 per year. Members receive newsletter.

Serial publications: *The Compendium.*

Service to libraries: Will answer reference questions by telephone or mail. Contact Ms. Nancy Pearlman. Brochures, fact sheets, petitions, and booklets available, some free and some for cost of postage.

Additional information: This group maintains a library, reads and reviews environmental impact reports, publishes lists of recycling centers and recycling companies, and "maintains the largest listing of ecology and conservation organizations in the Los Angeles and Orange Counties."

9-a-9 Environmental Action Coalition, Inc.
235 East 49th Street
New York, NY 10017
Telephone: 212-486-9550

Established: 1970

Priorities: Improvement in the quality of urban life. Current programs include the Recycling Program and the Educational Services Program.

Institutional membership: Open to libraries. $10 per year. Members receive subscription to *Eco-News* with accompanying teacher's guides; *Cycle*, a membership newsletter; use of library; loan of films.

Serial publications: *Eco-News*, monthly newsletter for children.

Services to libraries: Will answer reference questions by mail only. Contact Terry S. Ratcliff. Vertical file material on solid waste and environmental education (elementary level) is available free or at cost of postage.

9-a-10 Environmental Action, Inc.
Room 731, 1346 Connecticut Avenue NW
Washington, DC 20036
Telephone: 202-833-1845

Established: 1970

Priorities: Environmental education. Environmental legislation, particularly that concerned with air pollution, solid waste, transportation, and land use.

Institutional membership: E. A. has "supporters," rather than members; subscription to the magazine is $10; minimum supporter contribution is $15.

Serial publications: *Environmental Action*, biweekly.

Services to libraries: Will answer reference questions by telephone or mail. Latter should be addressed to Mailing Dept.

9-a-11 Environmental Action Foundation, Inc.
724 Dupont Circle Building
Washington, DC 20036
Telephone: 202-659-9682

Established: 1970

Priorities: Research and education on the environment. Current projects include: investigation of the environmental and social effects of electric utilities; research on the environmental impact of the B-1 bomber; assistance to citizens who are planning balanced transportation systems; citizen education on solid waste; and education on nuclear energy issues.

Institutional membership: Not a membership organization.

Services to libraries: Staff will answer reference questions by telephone or mail on subjects in their fields of expertise.

9-a-12 Environmental Defense Fund
162 Old Town Road
East Setauket, NY 11733
Telephone: 516-751-5191

Established: 1967

Priorities: Water resource management, energy policy, pest control, environmental health, land use and transportation.

Institutional membership: Open to libraries. $15 per year.

Serial publications: *EDF Letter*, three or four times a year.

Services to libraries: Will answer reference questions, preferably by mail. Contact Norma H. Watson. *EDF Letter* available free upon request. Will provide case summaries upon request, free for single copies, 5¢ per page for multiples. Also has available names of environmentally active lawyers, for groups wishing to litigate.

Additional information: "We are not a good source of information on general environmental questions although we do provide listings of sources of this information. We can provide information on environmental law and, through our Case Summaries, applications of laws. Other than our legal briefs and scientific testimony, we are not a source of publications."

Regional and special offices:

1525 18th Street NW
Washington, DC 20036

1130 Capitol Life Center
Denver, CO 80203

2728 Durant Avenue
Berkeley, CA 94704

Program Support
527 Madison Avenue
Suite 1217
New York, NY 10022

9-a-13 Friends of the Earth
529 Commercial Street
San Francisco, CA 94111
Telephone: 415-391-4270

Established: 1969

Priorities: Preservation, restoration and more rational use of the earth. Program is fourfold: conservation lobbying, litigation, publishing, and organizing worldwide.

Institutional membership: Open to libraries. $20 per year. Members receive subscription to *Not Man Apart*, fortnightly newspaper.

Services to libraries: Will answer reference questions by telephone (preferably members only) or mail. Will provide one copy of *Not Man Apart* and pertinent data as available for cost of postage.

Additional information: *Not Man Apart* is in the process of being indexed and the index will be available soon.

Regional and special offices: Fairbanks, AK; Tempe, AZ; Arcata, CA; Manhattan Beach, CA; Evergreen, CO; Lombard, IL; Cambridge, MA; Clinton, MO; Columbia, MO; Billings, MT; New York, NY; Philadelphia, PA; Madison, WI; Seattle, WA; and Washington, DC.

9-a-14 Environmental Education Group
6731 Lurline Avenue
Canoga Park, CA 91306
Telephone: 213-340-7309

Established: 1972 (evolved out of Environmental Action Group)

Priorities: Environmental education programs; research and reporting; consultation; prototypic projects.

Institutional membership: Not a membership organization.

Serial publications: *Pollution Watch*, monthly newsletter.

Services to libraries: Will answer reference questions by mail only. Vertical file material available, some free, some for postage, and some priced. Also a source of slides, charts, and other nonprint materials.

Additional information: "We are a clearinghouse and original resource for energy information."

9-a-15 International Union for the Conservation of Nature and Natural Resources
1110 Morges
Switzerland
Telephone: 021-71-4401

Established: 1948

Priorities: Maintaining and enhancing the diversity of the biosphere by promoting rational management of the earth's resources.

Institutional membership: Open to libraries. However, "in this particular case (libraries), there is no advantage in joining IUCN. Membership fees are $50 for an affiliate member."

Serial publications: *IUCN Bulletin*, monthly; *IUCN Yearbook*, annual.

Services to libraries: Will answer reference questions by mail only. Contact librarian. The IUCN maintains a special library which outsiders and students may consult on appointment. Xerox copy service is available.

Additional information: The IUCN publishes the *Red Data Book*, loose-leaf volumes listing information on rare and endangered species. New and replacement sheets are issued from time to time.

9-a-16 Izaak Walton League of America
1800 North Kent Street, Suite 806
Arlington, VA 22209
Telephone: 703-528-1818

Established: 1922

Priorities: Protection and restoration of America's outdoors and conservation of our natural resources through public education, policy research, citizen involvement, and legal action.

Institutional membership: Not open to libraries.

Serial publications: *Outdoor America*, monthly.

Services to libraries: Will answer reference questions by telephone or mail "when time permits." Provides citizen-action-oriented information, free or for cost of postage.

9-a-17 League of Conservation Voters
620 C Street SE
Washington, DC 20003
Telephone: 202-543-4312

Priorities: Support environmental candidates who face especially close races.

Institutional membership: Not a membership organization.

Services to libraries: Will answer brief reference questions by telephone or mail. Publishes voting charts showing how representatives and senators voted on key environmental issues ($1 each).

9-a-18 League of Women Voters of the United States
1730 M Street NW
Washington, DC 20036
Telephone: 202-296-1770

Established: 1920

Priorities: Citizen education to encourage participation in government, political action in the public interest. In environmental issues: air, water, solid waste, energy.

Institutional membership: Not open to libraries.

Services to libraries: Will answer reference questions limited to the League's environmental activity by telephone or mail. Contact the Environmental Quality Department.

9-a-19 National Audubon Society, Inc.
950 Third Avenue
New York, NY 10022
Telephone: 212-832-3200

Established: 1905

Priorities: Conservation of wildlife and the natural environment; ornithology; educating the public regarding its role within the natural environment.

Institutional membership: Open to libraries. $13 per year includes subscription to *Audubon Magazine*.

Serial publications: *Audubon Magazine*, bimonthly; *Audubon Leader Conservation Newsletter*, twenty-three times a year; *American Birds*, bimonthly.

Services to libraries: Will answer reference questions by telephone (ask for extension 270) or mail. Contact Mrs. Nancy T. Manson. Vertical file material on environmental issues available free upon request.

Regional and special offices: Harwinton, CT; Newfoundland, NJ; Atlanta, GA; Owensboro, KY; Red Wing, MN; Alma, KS; Austin, TX; Boulder, CO; Sacramento, CA.

9-a-20 National Environmental Health Association
1600 Pennsylvania Street
Denver, CO 80203
Telephone: 303-832-1550 (and 832-1601)

Established: 1939

Priorities: An organization of professionally trained men and women working to control environmental hazards and permit attainment of highest possible human health standards.

Institutional membership: Open to libraries as agency members. $35 per year. Members receive subscription to *Journal of Environmental Health*.

Serial publications: *Journal of Environmental Health.*

Services to libraries: Will answer reference questions by telephone or mail. Contact Nicholas Pohlit. Career information available free upon request (in "reasonable amounts").

9-a-21 National Parks and Conservation Association
1701 18th Street NW
Washington, DC 20009
Telephone: 202-265-2717

Established: 1919

Priorities: Protection of the great national parks and monuments of America, as well as conservation and restoration of the natural environment in general.

Institutional membership: Open to libraries. $12 per year. Members receive monthly magazine.

Serial Publications: *National Parks and Conservation Association Magazine: The Environmental Journal*, monthly.

Services to libraries: Will answer reference questions by mail. Contact Mrs. Adelaide MacLeod. Limited supply of technical position papers, some free and some priced.

9-a-22 National Wildlife Federation
1412 16th Street NW
Washington, DC 20036
Telephone: 202-483-1550

Established: 1936

Priorities: Seeks solutions to problems confronting man, wildlife, and the environment.

Institutional membership. Open to libraries. $7.50 per year; includes subscription to *National Wildlife*. $12.50 per year; includes *International Wildlife*. All with discounts on books and records.

Serial publications: *National Wildlife*, bimonthly; *International Wildlife*, bimonthly; *Conservation News*, semimonthly newsletter; *Conservation Report*, weekly; *Ranger Rick's Nature Magazine*, monthly.

Services to libraries: Will answer reference questions by telephone or mail. Contact library for either. Vertical file material on ecology and conservation available free upon request.

9-a-23 Natural Resources Defense Council, Inc.
15 West 44th Street
New York, NY 10036
Telephone: 212-869-0150

Established: 1970

Priorities: Uses legal and scientific techniques, including monitoring of major national legislation, to approach issues such as air, water, and noise pollution, energy, land use, and natural resource utilization.

Institutional membership: Open to libraries. $10. Members receive quarterly newsletter.

Serial publications: Newsletter, quarterly.

Services to libraries: Will answer reference questions by telephone or mail. Contact Marc Reisner or Carol Hine. Citizen guide books available for cost of postage.

Regional and special offices:

917 15th Street, NW
Washington, DC 20005
202-737-5000

664 Hamilton Avenue
Palo Alto, CA 94301
415-327-1080

9-a-24 The Nature Conservancy
1800 North Kent Street
Arlington, VA 22209
Telephone: 703-524-3151

Established: 1951

Priorities: Preservation of ecologically and environmentally significant land.

Institutional membership: Open to libraries. $10 (minimum) per year. Members receive *The Nature Conservancy News*.

Serial publications: *The Nature Conservancy News*, quarterly.

Services to libraries: Will answer reference questions by telephone or mail. Contact Public Relations Department. Some vertical file material free upon request.

Regional and special offices:

294 Washington Street
Boston, MA 02108
617-542-1908

325 W. 15th Street
Minneapolis, MN 55403
612-339-7231

4285 Memorial Drive
Decatur, GA 30032
404-294-7054

425 Bush Street
San Francisco, CA 94108
415-989-3056

9-a-25 Planned Parenthood—World Population
810 Seventh Avenue
New York, NY 10019
Telephone: 212-541-7800

Priorities: Voluntary fertility control; achieving U.S. population of stable size in an optimum environment; relevant research, information, education, and training programs.

Serial publications: *Family Planning Perspectives*, quarterly; *Current Literature in Family Planning*, monthly annotated bibliography.

Services to libraries: None. Will *not* answer reference questions.

9-a-26 Population Institute
110 Maryland Avenue NE
Washington, DC 20002
Telephone: 202-544-3310

Established: 1969

Priorities: Enlist and assist key leadership groups and leaders in developing innovative population-related activities.

Institutional membership: Not a membership organization.

Services to libraries: Will answer reference questions by mail only. Contact Yolande V. Baker.

Additional information: The Youth and Student Division of the Population Institute serves as a resource/clearinghouse gathering information on population issues and sharing it with high-school and college students.

Regional and special offices:

Youth & Student Division
Population Institute
110 Maryland Avenue, NW
Washington, DC 20002
202-544-3310

9-a-27 Population Reference Bureau, Inc.
1755 Massachusetts Avenue NW
Washington, DC 20036
Telephone: 202-232-2288

Established: 1929

Priorities: Devoted to gathering, interpreting, and publishing information on population dynamics and their implications for social and economic development and the environment.

Institution membership: Open to libraries. $10. Members receive *Population Bulletin*, *PRB Reports*, and *World Population Data Sheet*.

Serial publications: *Population Bulletin*, bimonthly; *PRB Reports*, periodically; *World Population Data Sheet*, annual.

Services to libraries: Will answer reference questions by telephone or mail. Contact Frances B. Jacobson, Librarian, extension 44. Sample copy of *Bulletin* or *Data Sheet* free upon request.

Additional information: *PRB Reports* contain an annotated bibliography, "Current Readings in Population."

9-a-28 Rachel Carson Trust for the Living Environment, Inc.
8940 Jones Mill Road
Washington, DC 20015
Telephone: 301-652-1877

Established: 1965

Priorities: Chemical contamination of the environment, particularly pesticides.

Institutional membership: Not a membership organization. Subscription to all publications is $10 per year.

Serial publications: *Down the Drain*, a series on water quality and preventive measures.

Services to libraries: Will answer reference questions by telephone ("if we can") or mail (for subscribers, within reason). Contact the executive director. A list of current publications will be sent to librarians on request.

9-a-29 Resources for the Future, Inc.
1755 Massachusetts Avenue NW
Washington, DC 20036
Telephone: 202-462-4400

Established: 1952

Priorities: Research on environmental problems.

Institutional membership: Not a membership organization.

Serial publications: *Resources*, newsletter, three times a year.

Services to libraries: Will *not* answer reference questions. Libraries may be put on mailing list to receive free newsletter.

9-a-30 Scientists' Institute for Public Information
30 East 68th Street
New York, NY 10021
Telephone: 212-249-3200

Established: 1963

Priorities: Clearinghouse for scientific information related to public policy issues such as occupational health and safety, energy, air, water and soil pollution, effects of pesticides, effects of radiation, lead, mercury, and other heavy metals.

Institutional membership: Open to libraries. $25 per year. Members receive *Environment* magazine and *SIPI Letter*.

Serial publications: *Environment*, monthly; *SIPI Letter*, quarterly.

Services to libraries: Will answer reference questions by telephone or mail. Contact Sally Brainin, Harriet Hotzman, or Nora Adelman. Some vertical file material available free upon request. Speakers bureau available. Scientific information library open to the public.

9-a-31 Sierra Club
1050 Mills Tower
San Francisco, CA 94104
Telephone: 415-981-8634

Established: 1892

Priorities: Environmental conservation, with special emphasis on energy conservation and protection of public lands.

Institutional membership. Open to libraries. $25 per year to be Friend of the Sierra Club. Requires written endorsement of purposes of the Sierra Club. Friends receive all publications normally sent to members.

Serial publications: *Sierra Club Bulletin*, monthly; *National News Report*, weekly; *International Report*, biweekly.

Services to libraries: Will answer reference questions by telephone (unless extensive information is requested) or mail. Contact Christie Hakim, Librarian, Sierra Club Library. Vertical file material available free, but donation requested. Priced publication list available.

Regional and special offices:

Sierra Club Foundation Headquarters
220 Bush Street
San Francisco, CA 94104
415-981-8637

Sierra Club Legal Defense Fund Headquarters
311 California Street, Suite 311
San Francisco, CA 94104
415-398-1411

Offices in: Anchorage, AK; New York, NY; Madison, WI; Seattle, WA; Sacramento, CA; Los Angeles, CA; Tucson, AZ; Washington, DC; Dubois, WY; Denver, CO.

Chapters in: Los Angeles, CA; New York, NY; Chicago, IL; Honolulu, HI; Palo Alto, CA; Lansing, MI; Sacramento, CA; Boston, MA; Princeton, NJ; Santa Fe, NM; Boulder, CO; San Diego, CA; Oakland, CA; Toronto, Ontario, Canada.

9-c-4 Environmental Health Institute
Pharmacy Building
Purdue University
West Lafayette, IN 47907
Telephone: 317-494-8537

Established: 1966

Priorities: Environmental toxicology and environmental radiological health.

Services to libraries: Will answer reference questions by mail only. Reprints of papers available free upon request.

9-c-5 Environmental Health Research & Training Center
University of Minnesota
Minneapolis, MN 55455
Telephone: 612-373-8086

Established: 1966

Priorities: Overall relationships of environmental contaminants to health.

Services to libraries: Will answer reference questions by telephone or mail. Call 616-373-8080. Contact Conrad P. Straub, Director.

9-c-6 Environmental Health Surveillance Center
Box 199
Route 4
Columbia, MO 65201
Telephone: 314-449-3947

Established: 1968

Priorities: Births, deaths, morbidity in geographically defined areas within the state in relationship to the local environmental deficiencies.

Services to libraries: Will answer reference questions by mail only. Reprints available free upon request.

9-c-7 Environmental Information Analysis Center
Battelle
Columbus Laboratories
505 King Avenue
Columbus, OH 43201
Telephone: 614-299-3151 ext. 1185

Established: 1964

Priorities: Supports the varied research programs conducted by the Ecology and Ecosystems Analysis Section. Broad areas of coverage include: management of environmental systems; water quality management; thermal effluent studies; ecosystem simulation; ecological studies; nuclear reactors; chemistry and technology; radioisotope applications; solid waste management; mineral, metallurgical, and metals finishing waste control; carbonaceous industrial waste control; food processing engineering; chemical and metal process technology; advanced process technology; environmental monitoring and analysis; particulate characterization; trace materials in the environment; materials characterization; monitoring and analysis methods; combustion research and equipment; air quality engineering; atmospheric chemistry; indoor environmental control; particle mechanics; incineration; environmental impacts assessment; environmental aspects of urban and regional planning; environmental elements of natural resource management; economic aspects of environmental planning; social and institutional aspects of environmental planning; environmental information management; recreational and aesthetic elements of environmental planning; land use planning; strip mining reclamation; and environmental aspects of urban development.

Publications: Bibliographies and data compilations.

Services to libraries: Will answer reference questions for *sponsors only*.

9-c-8 Environmental Physiology Laboratory
312 Hamilton Hall
1645 Neil Avenue
Columbus, OH 43210
Telephone: 614-422-7913

Established: 1940

Priorities: The gaseous environment; interaction between disease and components of the atmosphere.

Services to libraries: Will answer reference questions by telephone or mail, if possible. Contact H.S. Weiss. Reprints available free upon request.

9-c-9 Environmental Policy Center
Pennsylvania State University
401 Grange Building
University Park, PA 16802
Telephone: 814-865-1442

Established: 1972

Priorities: Assist the policy-setting agencies of the state and nation by providing information and analysis that is required to reach responsible resource management decisions.

Publications: Series of working papers.

Services to libraries: Will answer reference questions by telephone or mail. Contact Terry A. Ferrar, Director. Some vertical file material available free upon request.

9-c-10 Environmental Protection Research Institute, Inc.
24 Central Avenue
Waterbury, CT 06798
Telephone: 203-757-0787

Established: 1970

Priorities: Environmental information and technical data transfer, education, research, planning, analyses and conduct, international communications.

Publications: *Envirogram* (newsletter); *Journal of Environmental Research*, quarterly.

Services to libraries: Will answer reference questions by telephone "if possible." Call 203-757-0788. Will answer reference questions by mail. Contact Director, Resource Center and Library. Some vertical file material available free, some for cost of postage. The Institute also provides assistance to schools and interested organizations in designing and developing environmental curricula.

9-c-11 Environmental Studies Center
Bowling Green State University
Bowling Green, OH 43403
Telephone: 419-372-0207

Established: 1970

Priorities: Academic programs in environmental studies; consultation in vertebrate pest management; environmental evaluation; gaming and simulation; values clarification.

Publications: *EcoCentric*, newsletter for environmental educators; *Proceedings Bowling Green Bird Control Seminars*.

Services to libraries: Will answer reference questions "within reason" by telephone or mail. Contact the director.

9-c-12 Environmental Simulation Laboratory
2036 Dana Building
University of Michigan
Ann Arbor, MI 48104
Telephone: 313-763-1062

Established: 1967

Priorities: Gaming/simulation of environmental problems.

Services to libraries: Will answer reference questions by mail only. Reprints available for cost of postage.

9-c-13 Environmental Studies Center
11 Coburn Hall
University of Maine at Orono
Orono, ME 04473
Telephone: 207-581-7092

Established: 1970

Priorities: To encourage, coordinate, and conduct research and educational programs related to environmental concerns.

Publications: *Mainestream*, bimonthly; *Annual Report of Activities*; *Environmental Human Resource Directory*; *Municipal Guide for Shoreland Zoning*; *Water Quality and Recreational Land Use*; and a number of other monographs.

Services to libraries: Will answer reference questions by telephone or mail. Contact M.W. Hall. Vertical file materials are available, most free, some for cost of postage. Computerized reference library of mainly technical environmental material. Periodical listing of all publications, cassettes, brochures, and so forth, received by the Center.

9-c-14 Environmental Studies Institute
Baylor University
Waco, TX 76703
Telephone: 817-755-3406

Established: 1969

Priorities: Undergraduate and graduate education in environmental studies.

Services to libraries: Will answer reference questions by mail only. Contact Director. The Glassock Conservation Library is also available.

9-c-15 Environmental Trace Substances Center
412 Clark Hall
University of Missouri
Columbia, MO 65201
Telephone: 314-882-3321

Established: 1969

Priorities: Research on trace substances in environmental matrixes and analytical methodology; applied ecological research.

Publications: Annual proceedings of conference on trace substances in the environment.

Services to libraries: Will answer reference questions by telephone or mail. Contact the Director. Annual proceedings of conference available free upon request.

9-c-16 Hayes Research Foundation
PO Box 1404
Richmond, IN 47374
Telephone: 317-962-4894

Established: 1959

Priorities: Maintenance of complete Regional Arboretum of 178 native species of woody plants; education in ecology and natural history for all ages; community participation in outdoor experiences.

Publications: *Hayes Arboretum Calendar and Comment* (newsletter); *Welcome* (brochure); *Annual Fluctuations of Water Levels in Soils of the Miami Catena in Wayne County, Indiana* (Thorp & Gamble, 1972).

Services to libraries: Will answer reference questions by telephone (317-962-3745) or mail. Contact Don Hendricks, Associate Director. A fifteen-minute silent color motion picture *Discovery* (local nature scene) for free loan. Herbarium in field of interest; data on native (or naturalized) woody plants.

9-c-17 Indiana University—Aerospace Research Applications Center
Poplars Research & Conference Center
400 East 7th Street
Bloomington, IN 47401
Telephone: 812-337-7833

Established: 1963

Priorities: Provides technical assistance and information services to industry, government, and the public. Staff of scientists and engineers and a massive computer file of technical information.

Publications: Monthly and biweekly announcements of new technical literature and developments, including nearly two dozen topic areas in environment and energy-related fields.

Services to libraries: Provides background searches, current awareness services, and NASA document services, plus technology counseling and special contract activities for users.

Additional information: The center is a nonprofit unit of Indiana University that is supported by funds from NASA and fees for services provided to its users.

9-c-18 Information Center
Environmental Engineering Division
Texas A & M University
College Station, TX 77843
Telephone: 773-845-3011

Established: 1971

Priorities: Primary focus on water pollution, but also concerned with air pollution and solid waste.

Services to libraries: Will answer reference questions by mail. Contact Linda Clark.

Additional information: "The Information Center was designed to help professors within the Division in literature searching. Occasionally we do 'hire out' literature searching services if funds are provided by the requesting agency."

9-c-19 Institute of Environmental Stress
University of California
Santa Barbara, CA 93106
Telephone: 805-961-2350

Established: 1965

Priorities: To further fundamental research of an interdisciplinary nature towards an understanding of the adaptive potentials of organisms.

Services to libraries: Will answer reference questions by telephone or mail. Reprints available free upon request.

Additional information: Data bank present but not available for outside use.

9-c-20 Institute of Water Resources
Box U-37
University of Connecticut
Storrs, CT 06268
Telephone: 203-486-3534

Established: 1964

Priorities: Basic and applied research in the water resources field; information technology transfer.

Publications: Institute of Water Resources Report Series; *Connecticut Water Law: Summary and Index of Statutes.*

Services to libraries: Will answer reference questions by telephone or mail. Contact Dr. Victor E. Scottron, Director. Vertical file materials available, some free and some priced.

9-c-21 International Bird Rescue Research
2701 Eighth Street
Berkeley, CA 94710
Telephone: 415-841-9086

Established: 1971

Priorities: Rehabilitation of oil-coated aquatic birds and sea-bird physiology.

Publications: *International Bird Rescue Newsletter.*

Services to libraries: Will answer reference questions by telephone or mail. Contact David C. Smith or Alice B. Berkner. Reference library on wildlife health problems will soon be distributing microfiche card catalog.

9-c-22 Southwest Research Institute
8500 Culebra Road
San Antonio, TX 78284
Telephone: 512-684-5111

Established: 1947

Priorities: Nonprofit applied research.

Publications: *Tomorrow Through Research*, quarterly.

Services to libraries: Will answer brief reference questions by telephone (ext. 2126) or mail. Contact Edwin Vaught. *Tomorrow Through Research* is available free upon request.

9-c-23 Statewide Air Pollution Research Center
University of California
Riverside, CA 92502
Telephone: 714-787-5124

Established: 1961

Priorities: Development and conduct of both fundamental and applied research programs in key areas of air pollution, primarily: (1) on a wide range of problems in the plant sciences and (2) on the chemistry and physics of polluted atmospheres.

Publications: *California Air Environment*, three times a year, SAPRC occasional reports.

Services to libraries: Will *not* answer reference questions. Copy of *California Air Environment* free upon request.

9-c-24 Technical Guidance Center for Environmental Quality
317 Hills North
University of Massachusetts
Amherst, MA 01002
Telephone: 413-545-0347

Established: 1969

Priorities: An information and referral center for environmental problems and pollution abatement measures relative to water, air, solid waste disposal, soil, and noise; energy sources and conservation; natural resources development; land use; environmental control standards and legislation.

Publications: *TGC Bulletin—Reporting on the Environment.*

Services to libraries: Will answer reference questions by telephone or mail. Contact Ruth Kreplick, Director. *TGC Bulletin* available free upon request. The center also makes referrals to literature search sources.

Additional information: Emphasis is on public outreach program in the Commonwealth of Massachusetts; however, out-of-state responses are made as circumstances permit.

9-c-25 Toxicology Information Response Center
PO Box X, Building 7509
Information Center Complex
Oak Ridge National Laboratory
Oak Ridge, TN 37830
Telephone: 615-483-8611, ext. 3-1433

Established: 1971

Priorities: To collect, evaluate, and disseminate toxicology information on a wide variety of chemical classes: pharmaceuticals, industrial chemicals, food additives, pesticides, environmental pollutants.

Publications: Bibliographic reports on topical subjects with distribution through the National Technical Information Service, Springfield, Virginia.

Services to libraries: Will answer reference questions by telephone or mail. Contact Ms. Helga Gerstner, Coordinator. Brochure and fact sheets describing the Center, its services, policies, facilities and resources available free upon request.

On-line computer data bases MEDLINE and TOXLINE; Oak Ridge National Laboratory data bases.

Additional information: Resources and services are available to all individuals. Charges are assessed under a direct full-cost recovery system at the rate of $20 per hour for domestic search requests and $25 per hour for requests from other countries. *Short telephone inquiries and brief requests will be answered without charge.*

9-c-26 University of Notre Dame Environmental Research Center
Office of Advanced Studies
University of Notre Dame
Notre Dame, IN 46556
Telephone: 219-283-6291

Established: 1967

Priorities: Basic research on aquatic ecosystems, materials exchange in the land/water interface.

Services to libraries: Will answer reference questions by mail only. Contact the Director, UNDERC.

Additional information: The Environmental Research Center is located in Gobegic County, Michigan and Villas County, Wisconsin. It consists of six square miles of forested areas with over thirty bodies of water.

9-c-27 University of Pittsburgh — Pymatuning Laboratory of Ecology
Linesville, PA 16424
Telephone: 814-683-5813

Established: 1949

Priorities: Teaching and research in ecology, fresh water and terrestrial; faculty and graduate students carry out a year-round program of research in basic ecology, including studies of ecosystem energetics and population studies.

Publications: *The Pymatuning Symposia in Ecology* (volumes 1 through 4 have been published).

Services to libraries: Will answer reference questions by telephone (within limits) and mail. Contact Dr. Richard T. Hartman, Director. Brochure about summer program available free upon request. The Laboratory is part of the Northwestern Pennsylvania Data Bank with its headquarters at Allegheny College, Meadville, Pa.

Additional information: The Pymatuning Laboratory of Ecology maintains the C. A. Tryon Library as part of the Laboratory's facilities.

9-c-28 Water Resources Research Institute and Air Resources Center
Oregon State University
Corvallis, OR 97331
Telephone: 503-754-1022

Established: 1960

Priorities: Practical air and water problems.

Publications: Completion reports, seminar proceedings, newsletter, special reports.

Services to libraries: Will answer reference questions by telephone or mail. Some vertical file material available free upon request.

ENVIRONMENTAL CONSULTANTS

This listing of environmental consultants was prepared from a statistical sampling of those consultants listed in various standard reference sources. Every attempt has been made to insure that unique consulting firms have been identified and that there is an adequate geographical representation. The information, however, represents the responses of the consultants themselves and this listing should in no way be construed as a recommendation.

The questionnaire sent to the consultants requested the following information: name, address, telephone number, date established, description of the nature and scope of services, and additional information.

9-d-1 Alvord, Burdick and Howson
20 North Wacker Drive
Chicago, IL 60606
Telephone: 312-236-9147

Established: 1902

Nature and scope of services: "Consulting engineering services on the design of water supply and pollution control projects.

This includes feasibility studies, design reports, the preparation of plans and specifications, supervision of construction, and the preparation of operating and maintenance manuals."

Additional information: "The firm has specialized in environmental engineering for over seventy years; we have been consultants on the world's largest filtration plant for the City of Chicago and have served most of the large cities throughout the United States, including Chicago, Detroit, Denver, Fort Worth, Louisville, Roanoke, and many others."

9-d-2 Biometric Testing, Inc.
661 Palisade Avenue
Englewood Cliffs, NJ 07632
Telephone: 201-568-0224

Nature and scope of services: "Biometric Testing, Inc., consists of a group of creative researchers with experience in the health, food, and cosmetic sciences, including such diverse specialties as pharmacology, toxicology, environmental health, ecology, product development, clinical research, and governmental regulatory affairs. Our skills have been gained through personal experience in the design, performance and direction of research in biophysics, chemistry, medicine, food technology; in cosmetics chemistry; in industry; in the academic setting, the Food and Drug Administration, and the Atomic Energy Commission."

Additional information: "Additional office in Whippany, NJ."

9-d-3 Brandt Associates, Inc.
50 Blue Hen Drive
Newark, DE 19711
Telephone: 302-731-1550

Nature and scope of services: "Analysis of water and waste water."

Additional information: "Additional office: P.O. Box 81, Martins Creek, PA 18063."

9-d-4 Cambridge Acoustical Associates, Inc.
1033 Massachusetts Avenue
Cambridge, MA 02138
Telephone: 617-491-1421

Established: 1955

Nature and scope of services: "Consultants in noise control, environmental impact of noise. Architectural acoustics and underwater sound."

Additional information: "Our office published: Junger, Miguel C. and Feit, David, *Sound Structures, and Their Interaction* (Cambridge: MIT Press, 1972)."

9-d-5 Camp Dresser & McKee, Inc.
One Center Plaza
Boston, MA 02108
Telephone: 617-742-5151

Established: 1944

Nature and scope of services: "Camp Dresser & McKee firms specialize exclusively in the field of environmental engineering, involving improvements of land, air and/or water environments. Services offered include consultation, feasibility studies, surveys, investigations, master plans, reports, environmental planning, and environmental impact assessments. Typical facilities design projects include water supply, treatment, and distribution; flood control works; sewers, storm drainage, and waste water treatment; solid waste management systems, including incinerators, landfills and resource recovery; and area-wide and river basin planning studies."

Additional information: Additional offices in Atlanta, GA; Denver, CO; New York, NY; Pasadena, CA; Suitland, MD; Washington, DC; Asunción, Paraguay; Istanbul, Turkey; Lahore, Pakistan; Seoul, Republic of Korea; Singapore, Republic of Singapore; Victoria, Australia.

9-d-6 Cavanaugh Copley Associates
10 Bowers Street
Newton, MA 02160
Telephone: 617-965-5370

Established: 1970

Nature and scope of services: "Consulting in environmental acoustics to abate noise problems associated with highways, airports, industrial plants, and buildings in general."

9-d-7 Coastal Ecosystems Management, Inc.
3699 Hulen Street
Fort Worth, TX 76107
Telephone: 817-731-3727

Established: 1970

Nature and scope of services: "General terrestrial and aquatic ecological baseline surveys, precise water quality analyses, land management, environmental statements and inventories, wildlife and forestry management, land planning, range and grassland studies. Offshore environmental parameter studies, ocean survey management, oceanographic instrumentation, deep submergence engineering. Sedimentological studies, water quality and ecosystem modeling, and general geology."

Additional information: "We have distributed over five hundred copies of printed or multilithed reports on environmental studies of terrestrial and aquatic ecosystems. Several of our reports are proprietary or confidential and permission to cite must be obtained from our clients. Certain reports can be obtained from C.E.M. upon request. If printed copies are unavailable, copies can be made at 10¢ per page. Our environmental library is listed by U.S. Environmental Protection Agency in their directory of state and local environmental libraries (July 1973)."

9-d-8 Cole Associates, Inc.
3600 East Jefferson Boulevard
South Bend, IN 46615
Telephone: 219-288-9131

Established: 1916

Nature and scope of services: "Complete engineering and architectural services for business, industry and government. Special services in the areas of environmental problems, environmental impact statements, studies, feasibility studies, and so forth."

Additional information: "We will be pleased to discuss environment problems of persons or groups referred to us. Speaker service may be arranged."

9-d-9 Commonwealth Associates Inc.
209 East Washington Avenue
Jackson, MI 49201
Telephone: 517-787-6000

Established: 1920

Nature and scope of services: "Engineering, architecture, and planning. Employ 1,500 people, international in service, secondary offices in South America and Iran."

Additional information: "Commonwealth is a subsidiary company of Gilbert Associates, Reading, PA. The two companies employ about 5,000 people providing only design services, primarily to the utility industry (power plants, transmission, substations).

"Over fifty environmental reports have been prepared by Commonwealth for the utility industry. Two probably are most important: EHV transmission environmental guidelines for Minnesota [and] EHV transmission guidelines for the northeast power coordinating committee."

9-d-10 Compact Air Samplers
825 Belmonte Park North
Dayton, OH 45405
Telephone: 513-278-3891

Established: 1955

Nature and scope of services: "Consulting services and laboratory services in the field of occupational health, air and water pollution."

9-d-11 Controls for Environmental Pollution, Inc.
PO Box 5351
Santa Fe, NM 87501
Telephone: 505-982-9841

Established: 1970

Nature and scope of services: "Consulting and laboratory service providing a complete 'package' to industries and other concerns with a need for monitoring possible sources of pollution. Specialists in developing monitoring programs for nuclear-power generating companies."

Additional information: "Affiliated with Diversified Biological

Systems, San Antonio, TX; and Ludlum Measurements, Inc., Sweetwater, TX; and Todd Research and Technical Division, Galveston, TX, for special program requirements."

9-d-12 Leo A. Daly Co.
8600 Indian Hills Drive
Omaha, NE 68114
Telephone: 402-391-8111

Established: 1915

Nature and scope of services: "The Daly organization provides architectual, engineering, and planning services, as they relate to both the urban and regional settings. We are particularly cognizant concerning the relationship of 'energy management,' environmental considerations, and the comprehensive planning and/or detailed designs accomplished by our organization. This we accomplish by means of multidisciplinary teams of architects; landscape architects, interior designers; environmentalists; systems, environmental, civil, structural, mechanical, and electrical engineers; and comprehensive, master, and site planners."

Additional information: "The Daly organization has extensive experience in the design of libraries. Additional offices in Los Angeles, St. Louis, San Francisco, Seattle, Washington, DC, Hong Kong, Singapore, and Jakarta, Indonesia."

9-d-13 Demopulos and Ferguson, Inc.
600 Petroleum Tower
Shreveport, LA 71101
Telephone: 318-221-7117

Established: 1953

Nature and scope of services: "Complete civil and structural engineering services including preliminary planning, feasibility studies, design, contract plans; specifications, and construction inspection of highways, bridges, airports, water supply, treatment, and distribution, solid waste disposal, industrial waste and sewage treatment, drainage, flood control, and land planning for military installations, industrial, commercial and residential developments."

9-a-32 Student Conservation Association, Inc.
Route 1, Box 573A
Olympic View Drive
Vashon, WA 98070
Telephone: 206-567-4798

Established: 1964

Priorities: Conducts youth volunteer service programs in national parks and forests.

Institutional membership: Any amount donation. Members receive annual mailing of program announcement.

Services to libraries: Does *not* answer reference questions. Annual list of volunteer positions available free upon request.

9-a-33 Threshold: An International Center for Environmental Renewal
1785 Massachusetts Avenue NW
Suite 113
Washington, DC 20036
Telephone: 202-265-0020

Established: 1972

Priorities: Demonstration projects, ecological land-use planning, ecological technology, national park planning, international impact assessment studies, environmental education.

Institutional membership: Open to libraries. Associate Status: $30. Benefactors: $100 and up.

Service to libraries: Will answer reference questions by telephone or mail. Contact John P. Milton, Peter H. Freeman, or James Aldrich.

9-a-34 Water Pollution Control Federation
3900 Wisconsin Avenue NW
Washington, DC 20016
Telephone: 202-537-1320

Established: 1928

Priorities: The advancement of fundamental and practical knowledge about waterborne wastes and waterborne waste systems, treatment, and reclamation works.

Institutional membership: Open to libraries. May join as Corporate Members for annual dues ranging from $77 to $100; may attend conferences, receive staff assistance or reference materials, and participate in training and education programs.

Services to libraries: Will answer reference questions, preferably by mail.

9-a-35 The Wilderness Society
1901 Pennsylvania Avenue NW
Washington, DC 20006
Telephone: 202-293-2732

Established: 1935

Priorities: Wilderness preservation and management, land use, timber, management, strip mining, resource development, wildlife and habitat, energy development, dams, stream channelization and diversion.

Institutional membership: Open to libraries. $5. Members receive all regular membership publications, including *The Living Wilderness*, *Wilderness Report*, and *Conservation Alert*.

Serial Publications: *The Living Wilderness*, quarterly; *Wilderness Report*, quarterly newsletter; *Conservation Alert*, irregular.

Services to libraries: Will answer reference questions on wilderness issues by telephone or mail. Back issues of Wilderness Society publications free upon request (if available).

Regional and special offices:

4260 East Evans Avenue
Denver, CO 80222
303-758-2266

9-a-36 The Wildlife Society
3900 Wisconsin Avenue NW
Washington, DC 20016
Telephone: 202-363-2435

Established: 1936

Priorities: Establish and maintain high professional standards in wildlife management.

Institutional membership: Open to libraries. $40 as subscribing member.

Serial publications: *The Journal of Wildlife Management*, quarterly; *Wildlife Society Bulletin*, bimonthly.

Services to libraries: Will *not* answer reference questions.

9-a-37 World Wildlife Fund—U.S.
910 Seventeenth Street NW
Suite 619
Washington, DC 20006
Telephone: 202-296-0422

Established: 1961

Priorities: An international conservation organization represented in twenty-five countries, including the United States. Dedicated to preserving the world's endangered wildlife and natural areas.

Institutional membership: Not a membership or organization.

Serial publications: *Special Progress Report*, biannual.

Service to libraries: Will answer reference questions, preferably by mail. Contact information office. Endangered species list and informational brochures are available free (or at cost of postage when large amounts requested).

9-a-38 Zero Population Growth
1346 Connecticut Avenue NW
Washington, DC 20036
Telephone: 202-785-0100

Established: 1968

Priorities: Stabilization of U.S. population size, followed by reduction to a more reasonable level; reduction in the rate of growth, and eventual stabilization of U.S. consumption of non-renewable resources; rapid stabilization of U.S. energy consumption; land-use planning at all levels of government that will assure the stewardship of this limited resource.

Institutional membership: Open to libraries. $5.50 to subscribe to the *ZPG National Reporter*; $15 to become a member, which includes receiving the *ZPG National Reporter*.

Serial Publications: *ZPG National Reporter*.

Services to libraries: Will *not* answer reference questions. Provides vertical file material about population and other growth questions free upon request. Also provides teacher's materials, local growth information, media assistance, state legislation information.

LIBRARIES

Most large research collections include extensive holdings in the field of ecology. This listing of libraries only identifies those which are devoted solely to collecting material on some aspect of the environment and/or are prepared to service the needs of other libraries. Many research libraries undoubtedly have collections as strong as those listed here. Our purpose in listing these libraries, however, is to provide librarians with a starting point in supplementing their own collections.

The notation (EPA) following the name of a library indicates that the library is operated by the U.S. Environmental Protection Agency.

The questionnaire sent to libraries requested the following information: name, address, telephone number, date established, statement on scope of collection, publications, services rendered to other libraries, and additional information.

9-b-1 Academy of Natural Sciences of Philadelphia Library
19th and Benjamin Franklin Parkway
Philadelphia, PA 19103
Telephone: 215-107-3700

Established: 1812

Scope of collection: Systematic biology, palaeontology, ecology, descriptive geology. Approximately 150,000 volumes and 3,000 periodicals. Manuscript collection contains more than 200,000 items.

Publications: *Catalog of the Library of the Academy of Natural Sciences of Philadelphia* (G.K. Hall, 1972); *Guide to the Manu-*

script Collections in the Academy of Natural Sciences of Philadelphia (The Academy, 1963).

Services to other libraries: Will occasionally answer by mail questions specifically regarding the collection. Photocopies provided for a fee. Interlibrary loan according to code. Open to the public for reference only.

9-b-2 Air Pollution Technical Information Center (EPA)
Research Triangle Park
NC 27711
Telephone: 919-549-8411

Established: 1966

Scope of collection: Basic data and other information related to air pollution as reported in worldwide scientific and technical literature. Coverage includes 7,000 periodicals, reports, patents, dissertations, translations, technical papers, and articles from books and proceedings.

Publications: *Air Pollution Abstracts*, monthly; *Air Pollution Technical Publications of the U.S. Environmental Protection Agency*, semiannual; bibliographies.

Services to other libraries: Will answer reference questions by telephone or mail. Vertical file material free upon request. Data bank (70,000 abstracts) provides free retrospective retrievals. Retrievals of citations only at ten EPA regional office libraries. (See 9-b-15 to 9-b-19, for the addresses of some of these libraries.) All abstracts are in English.

9-b-3 Atmospheric Sciences Library
8060 Thirteenth Street, Room 806
Silver Spring, MD 20910
Telephone: 301-427-7800

Scope of collection: Approximately 150,000 volumes in the fields of meteorology, hydrology, climatology, and related sciences. Extensive collections of daily weather maps and observations and official meteorological publications of the United States and foreign governments.

Publications: Library accessions list.

Services to other libraries: Will answer reference questions by telephone or mail. Access to OASIS (Oceanic and Atmospheric Scientific Information System), developed by the National Oceanic and Atmospheric Administration (Department of Commerce).

9-b-4 Battelle Memorial Institute Library
505 King Avenue
Columbus, OH 43201
Telephone: 614-299-3151

Established: 1929

Scope of collection: Scientific and technical information with emphasis on chemistry, physics, materials, and metallurgy.

Publications: *BCL Serials List* ($25).

Services to other libraries: Will answer reference questions by telephone (ext. 2213) or mail. Contact Miss Audrey Jackson. Provides vertical file material concerned with Battelle activities (write to Publications Department). Participates in interlibrary loan; provides photo copy service for $1.50 handling charge plus 10¢ per page. Energy Information Center data bank can be searched for a fee. Other data bases, such as Lockheed, are rented by Battelle.

9-b-5 Conservation Library Center
1357 Broadway
Denver, CO 80203
Telephone: 303-573-5152 ext. 254, 262

Established: 1960

Scope of collection: Field of environment (conservation), including ecology, pollution, fish, wildlife, and land use, as well as the economic, social, political, historical, technical, and managerial aspects of the environment. International in scope but stress is on the United States.

Publications: *Catalog of the Conservation Library* (published by G.K. Hall); *Series on Unique and Endangered Species*; *Selected New Readings in Conservation*; *Environmental Education*; bibliographies on various topics.

Services to other libraries: Will answer reference questions by telephone (if not too involved) or mail. If the request requires extended staff time, the person must go to the library or the library must charge for time spent on the question. Contact Ms. Kay Collins. Some vertical file material, generally printed bibliographies or brochures, available free upon request. The library center also provides interlibrary loan, photocopy at cost, search of selected data base on fish and wildlife literature for $30 per question, and access to water data base.

Additional information: "Will make referrals to other known sources. If coming from out of town, please make appointment to insure material is available."

9-b-6 Environmental Conservation Library of Minnesota (ECOL)
Minneapolis Public Library and Information Center
300 Nicollet Mall
Minneapolis, MN 55401
Telephone: 612-372-6609

Scope of collection: Basic collection of general environmental materials and in-depth coverage of natural resources and environment of the Upper Midwest. Both popular and scholarly works are included.

Publications: *ECOL Book Catalog* (See 3-a-7); *ECOL News*, quarterly newsletter.

Services to other libraries: Will answer reference questions by telephone or mail. Brochures describing the library free upon request. Some bibliographies available for small charge.

Additional information: The library maintains a collection of environmental impact statements related to Minnesota actions.

9-b-7 Field Studies Section Library (EPA)
PO Box 219
Wenatchee, WA 98801
Telephone: 509-663-0243
          509-663-0031 ext. 243

Established: 1952

Scope of collection: Pesticides research. Includes books, journals, reprints and material on poisonings, health hazards and safety precautions, published papers.

Services to other libraries: Will answer reference questions by

telephone or mail. Contact Mr. Homer R. Wolfe, Chief, Field Studies Section. Reprints available free upon request. Will make copies of requested articles from their holdings.

9-b-8 Illinois Institute for Environmental Quality Library
309 West Washington Street
Chicago, IL 60606
Telephone: 312-793-3870

Established: 1971

Scope of collection: Books, reports, journals, and microfiche on all aspects of the environment, including air pollution, water pollution, noise, solid waste, land use planning, thermal pollution, energy, environmental law.

Publications: *Catalog of Solid Waste Literature* (State of Illinois, 1973. Available through NTIS); *Catalog of Environmental Literature* (State of Illinois, 1974. Available through NTIS).

Services to other libraries: Will answer reference questions by telephone or mail. Contact Anela Trabert or Betty Ladner. Limited supplies of vertical file material available free upon request.

Additional information: Library is open to the general public. Most items may be borrowed for a limited time. Library also has a computer terminal which allows access to information about current bills before the Illinois legislature.

9-b-9 National Environmental Research Center (EPA)
200 SW 35th Street
Corvallis, OR 97330
Telephone: 503-752-4211 ext. 346, 421

Established: 1966

Scope of collection: Technical and scientific collection on pollution with an emphasis on ecological effects rather than control of pollution. Books and journals selected to support research on eutrophication, thermal pollution, coastal pollution, industrial pollution (including food wastes and pulp and paper wastes) and the effects of air pollution on plants and animals.

Publications: *Publications and Reports of the Pacific Northwest Environmental Research Laboratory, Corvallis, Oregon*, revised Spring 1974.

Services to other libraries: Will answer reference questions by

telephone or mail. "However, since the collection is pretty technical for most public library questions we would suggest contacting the Public Affairs Office first. They answer lay level questions and offer free pamphlets and brochures. Call 503-752-4211 ext. 300."

Additional information: The Office of Public Affairs publishes a series of twenty-one research-highlight newsletters of a semitechnical nature.

9-b-10 National Institute of Environmental Health Sciences Library
PO Box 12233
Research Triangle Park
NC 27709
Telephone: 919-549-8411 ext. 3426

Established: 1967

Scope of collection: Collection of 225 mission-oriented journals (environmental research) and approximately 3,000 books, including serials.

Publications: *Environmental Health Perspectives* (journal available through Government Printing Office).

Services to other libraries: Will answer reference questions by telephone or mail. Contact Ralph Hester. Limited supply of annual report available free upon request. Bibliographic services available upon request in response to biochemical and biomedical inquiries. The library subscribes to several data bases, such as MEDLINES, MEDLARS, and SDC, for On-Line services.

Additional information: Few subscriptions date back beyond 1967.

9-b-11 National Water Quality Laboratory Library (EPA)
6201 Congdon Boulevard
Duluth, MN 55804
Telephone: 218-727-6692

Established: 1967

Scope of collection: Research library concentrating on fresh water and fish.

Services to other libraries: Will answer reference questions by

mail only. Contact Librarian, Mary Harden. Vertical file material available free upon request.

Additional information: "The general public is invited to use the NWQL library facilities for in-house research."

9-b-12 Oklahoma Environmental Information and Media Center
East Central University
Ada, OK 74820
Telephone: 405-332-8000 ext. 3088

Established: 1971

Scope of collection: Approximately 6,000 volumes hard copy and 15,000 microfiche; 150 newsletters and periodicals concerned with the environment.

Publications: *Eco Systems*, an environmental journal published eight times a year ($2 per year).

Services to other libraries: Will answer reference questions by telephone or mail "if possible."

Additional information: "We employ university students who are majoring in environmental science to do information searches manually. They can receive up to four college hour credits, also. Our searchers also use Linscheid Library at East Central University and the library of the nearby Robert S. Kerr Environmental Research Center, one of EPA's national labs."

9-b-13 Science/Technology Information Center
Clark Hall
University of Virginia
Charlottesville, VA 22901
Telephone: 804-924-3023

Established: 1971

Scope of collection: This collection was established to support undergraduate and graduate courses and faculty research in the environmental sciences. It relies for peripheral materials on other University of Virginia collections.

Services to other libraries: Will answer reference questions by telephone or mail. Contact Edwina Pancake.

Additional information: The Science/Technology Information

Center is moving into a new building during the summer of 1975. "The end result will incorporate all of the present Environmental Sciences Library in the Science/Technology Information Center with the larger staff of the Center, and hopefully, new services to our immediate users and perhaps also to users outside the [university] community."

9-b-14 Solid Waste Information Retrieval System Library (EPA)
1835 K Street NW, Room 631
Washington, DC 20406

Established: 1968

Scope of collection: Totally solid-waste oriented.

Services to other libraries: Will answer reference questions by mail only. Contact John Connolly. Vertical file material available free upon request.

Additional information: Solid Waste Information Retrieval Services (SWIRS), P.O. Box 2365, Rockville, MD 20852.

9-b-15 U.S. Environmental Protection Agency
National Field Investigation Center Library
Denver Federal Center
Box 25227
Denver, CO 80225
Telephone: 303-234-2122

Established: 1972

Scope of collection: The collection consists of technical reports which have been developed from surveys in specific geographic locations, industrial sites, and municipal plants, usually with water pollution enforcement as the basis for the study.

Publications: No published catalog. Substantial list of reports published by the Center.

Services to other libraries: Will answer reference questions by telephone or mail. Contact Shirley Engebrit. Vertical file material available free upon request.

9-b-16 U.S. Environmental Protection Agency
Region III Library
Curtis Building
6th and Walnut Street
Philadelphia, PA 19106
Telephone: 215-597-0580

Established: 1972

Scope of collection: Special collection on environmental education.

Publications: Holdings list on environmental education.

Services to other libraries: Will answer reference questions by telephone or mail. Contact Pauline Levin.

9-b-17 U.S. Environmental Protection Agency
Region IV Library
1421 Peachtree Street NE
Atlanta, GA 30309
Telephone: 404-526-5216

Established: 1973

Scope of collection: Subject areas covered in varying depth are water, air, solid waste, radiation, noise, and pesticide pollution, with a specific focus on these problems in the Southeastern region. There is some coverage of energy, land use, and management theory. The collection includes approximately 100 journal and newsletter titles, 400 books and 3,000 technical reports.

Services to other libraries: Will answer reference questions by telephone or mail. Vertical file material available free upon request.

Additional information: "We have a complete set of all EPA publications on microfiche."

9-b-18 U.S. Environmental Protection Agency
Region VII Library
1735 Baltimore
Kansas City, MO 64108
Telephone: 816-374-5828

Established: 1970

Scope of collection: Water pollution (largest subject), air pollution, solid waste, pesticides, and radiation.

Services to other libraries: Will answer reference questions by telephone or mail. Contact Connie McKenzie.

Additional information: Reports of congressional hearings from 1970 to present on microfiche (with indexes).

9-b-19 U.S. Environmental Protection Agency
Region VIII Library
1860 Lincoln Street
Denver, CO 80203
Telephone: 303-837-2560

Established: 1973

Scope of collection: The collection supports the Agency's mission in the fields of air, noise, pesticides, radiation, solid waste, and water. Emphasis is placed on the technical or legal aspects of these subjects.

Services to other libraries: Will answer reference questions by telephone or mail (brief answer only). Contact John Lattimer, Region VIII Librarian.

Additional information: The library contains large microfiche collections of technical reports which are concerned with air pollution or were done by the Agency.

9-b-20 U.S. Environmental Protection Agency
Region IX Library
100 California Street
San Francisco, CA 94111
Telephone: 415-556-1840

Established: 1969

Scope of collection: Environmental pollution, especially as it relates to California, Arizona, Hawaii, Nevada, and the Pacific Islands.

Publications: Acquisitions list.

Services to other libraries: Will answer reference questions by telephone (415-556-1841) or mail, within reason. Vertical file material available free upon request.

9-b-21 U.S. Environmental Protection Agency
Region X Library
1200 6th Avenue
Seattle, WA 98101
Telephone: 206-442-1289

Established: 1971

Scope of collection: All EPA reports with emphasis on Region X reports.

Publications: EPA 600/9-74-001 *Indexed Bibliography of Office of Research and Development Reports*; PB-223 693/3M *EPA Reports Bibliography*, July 1973; PB-234 215/2GA *EPA Reports Bibliography Supplement*, August 1974.

Service to other libraries: Will answer reference questions by telephone or mail.

9-b-22 University of Wisconsin—Green Bay, Library
Green Bay, WI 54302
Telephone: 414-465-2303

Established: 1969

Scope of collection: This library has been developed to serve a new university of about 4,000 students. The focus of the entire university is on the study of man in relation to his environment. Therefore, the collection has focused on environmental studies, including monographs, serials, government publications, and indexes and abstracts.

Publications: Lists and bibliographies produced from time to time.

Services to other libraries: Will answer reference questions by telephone or mail. Contact Reference Department. Occasional bibliographies available for cost of postage.

9-b-23 Washington State Department of Ecology Technical Library
Olympia, WA 98504
Telephone: 206-753-2959

Established: 1970

Scope of collection: Current and reference materials on environmental subjects as needed by department, with particular emphasis on Washington State.

Publications: Monthly list of new publications acquired.

Services to other libraries: Will answer reference questions by telephone. Only limited service by mail. ("Prefer other libraries contact Washington State Library, Olympia, Washington 98504.") Publications and lists of publications of the department available upon request. Access to Water Resources Information System on Washington State water resources.

RESEARCH ORGANIZATIONS

There are a number of organizations devoted to applied and theoretical research in the field of ecology. The listing which follows is highly selective and is derived from organizations' responses to the questionnaire summarized below. Through the standard references, we have attempted to identify unique research organizations. While not all these organizations are prepared to extend services to libraries, a number of them will, as is indicated here. All the organizations contribute to the fund of knowledge about key environmental problems and are, therefore, important sources of activity.

The questionnaire sent to research organizations requested the following information: name, address, telephone number, date established, priorities or focus, publications, services rendered to libraries, and additional information.

9-c-1 Bob and Bessie Welder Wildlife Foundation
PO Box 1400
Sinton, TX 78387
Telephone: 512-364-2643

Established: 1954

Priorities: Research into wildlife, wildlife–domestic livestock relationships, wildlife-range relationships, and related problems.

Services to libraries: Will answer reference questions by telephone or mail "when possible." Contact Mrs. Willa Glazener, librarian (part-time). Vertical file material available, some free, some for cost of postage.

9-c-2 Ecological Sciences Information Center
Building 2029, Box X
Oak Ridge National Laboratory
Oak Ridge, TN 37830
Telephone: 615-483-8611 ext. 3-6524 and 3-6915

Established: 1968

Priorities: Effects of power plants on aquatic systems; radio-ecology, environmental aspects of the transuranics, ecosystems analysis, fauna of Tennessee.

Publications: *Annual Reviews and Indexed Bibliographies on the Effects of Temperature on Aquatic Organisms; Indexed, Annotated Bibliography on Striped Bass; Indexed, Annotated Bibliographies on the Environmental Aspects of the Transuranics; Indexed, Annotated Bibliography on the ORNL Environmental Sciences Division Publications.*

Services to libraries: Will answer reference questions by telephone or mail. Contact Helen Pfuderer.

Additional information: All information is in an automated information file. Tabular data is available on plutonium in mammals.

9-c-3 Environmental Engineering Sciences Research Center
University of Florida
Gainesville, FL 32611
Telephone: 904-392-0834

Established: 1966

Priorities: Water supply, water pollution control, air pollution, radiological health, environment resources management, environmental biology, environmental chemistry, solid wastes, systems ecology.

Publications: *Ecotek*, annual newsletter; *Annual Research Report—Environmental Engineering Sciences.*

Services to libraries: Will answer brief reference questions by telephone or mail. Contact the chairman. Newsletter, brochure, and journal articles available free upon request.

9-d-14 Eco-Labs, Inc.
1836 Euclid Avenue
Cleveland, OH 44115
Telephone: 216-694-3740

Established: 1971

Nature and scope of services: "Eco-Labs, Inc., is an environmental, analytical, research laboratory located in Cleveland, OH. It was conceived, designed, and equipped to perform environmental analyses. It makes in-plant surveys, treatability studies, water quality surveys and performs routine analyses such as those required under the NPDES program of the U.S. Environmental Protection Agency."

Additional information: "In addition to analytical work we conduct literature surveys, develop computer simulation models of streams, and write training manuals."

9-d-15 Ecology Audits, Inc.
9995 Monroe Drive, Suite 107
Dallas, TX 75220
Telephone: 214-350-7893

Established: 1967

Nature and scope of services: "Ecology Audits, Inc. is a service organization providing services in source testing, ambient air monitoring, water testing, and the preparation of environmental impact statements and assessments."

Additional information: "Additional office in Lake Charles, LA."

9-d-16 Ekono, Inc.
410 Bellevue Way, SE
Bellevue, WA 98004
Telephone: 206-455-5969

Established: 1967

Nature and scope of services: "Consulting engineers who provide solutions to industrial pollution problems (air and water) through engineering teams who study internal changes in process as well as external treatment systems for effectiveness and minimum cost."

Additional information: "Also provide energy studies and process engineering studies with emphasis in the pulp and paper and forest products industry. Subsidiary of Ekono Oy, Helsinki, Finland."

9-d-17 Engineered Environments
The Anaconda Co., Brass Division
Box 747
Waterbury, CT 06720
Telephone: 203-757-2021

Established: 1970

Nature and scope of services: "Engineering and consulting services in the OSHA, noise pollution, water pollution, air pollution, and solid waste pollution fields. Special expertise is had in heavy metal chemistry and the nonferrous metals and textile finishing fields."

9-d-18 Environmental Quality Analysts, Inc.
428 Jessie Street
San Francisco, CA 94103
Telephone: 415-777-1070

Established: 1947

Nature and scope of services: "Analytical support to engineers designing water and wastewater treatment facilities."

Additional information: "Most information we have is proprietary and not available unless released by our clients.
"Additional office in Costa Mesa, CA."

9-d-19 Enviro-Pact Division of USC, Inc.
5100 Centre Avenue
Pittsburgh, PA 15232
Telephone: 412-687-4700

Established: 1973

Nature and scope of services: "The Enviro-Pact division of USC Incorporated is an organization of consultants qualified to assist in the preparation of environmental assessments and impact statements required by the National Environmental Policy Act (NEPA)."

Additional information: "Apt, Bramer, Conrad and Associates, under a new corporate reorganization, has been retained only for the purpose of registry to do business in the state of Illinois.

"Additional office in Falls Church, VA."

9-d-20 Calvin H. Gibson Associates, P.C.
380 Main Street
East Orange, NJ 07018
Telephone: 201-673-1850

Established: 1969

Nature and scope of services: "Professional engineers and land surveyors."

Additional information: "Consulting engineers to solid waste industry in field of resource recovery through reclamation and pyrolysis."

9-d-21 Grant, Brundage, Baker & Stauffer, Ltd.
4955 Arbor Village Drive
Columbus, OH 43214
Telephone: 614-888-3100

Established: 1937

Nature and scope of services: "Consulting sanitary engineers."

Additional information: "Additional office in Cincinnati, OH."

9-d-22 Holley, Kenney, Schott, Inc.
921 Penn Avenue
Pittsburgh, PA 15222
Telephone: 412-471-5348

Established: 1962

Nature and scope of services: "Engineering and construction services for the iron and steel, nonferrous, chemical, industrial minerals and coal industries. Capabilities include planning, feasibility studies, cost analysis, preliminary and detail design, construction and construction management for all aspects of environmental concerns."

Additional information: "Additional office in Beckley, WV."

9-d-23 Howard, Needles, Tammen and Bergendoff
1805 Grand Avenue
Kansas City, MO 64108
Telephone: 816-474-4900

Established: 1914

Nature and scope of services: "To meet the challenge of conducting environmental studies, including socio-economic analyses, the firm has assembled an interdisciplinary team of professionals trained in terrestrial and acquatic biology, forestry, environmental physics, economics, sociology, park and recreation planning, urban and regional planning, environmental engineering, and community relations. Experienced writers and editors, familiar with the most recent requirements of environmental legislation, function as an integral part of the study team."

Additional information: "HNTB's staff of more than 1,200 includes specialists in virtually all aspects of the planning-design-construction process for more than a dozen project types. The firm has pioneered in the administration of multidisciplinary design teams, and its decentralized structure, with twenty-seven domestic and four overseas offices, enables HNTB to offer the resources of a large national organization for solutions to local problems. The local office is in charge, organizing and assembling the design team, drawing on skills from other offices as the job requires, supervising the program, and providing direct, personalized client service at all stages.

"Additional office in Alexandria, VA."

9-d-24 Industrial Noise Services
543 Bryant Street
Palo Alto, CA 94301
Telephone: 415-321-7911

Established: 1971

Nature and scope of services: "Consulting engineering in noise control for industry and providing yearly hearing tests to workers exposed to industrial noise."

Additional information: "Additional offices in Norwalk, CA; Lutherville, MD; Houston, TX; Hinsdale, IL; Springfield, PA; New York, NY."

9-d-25 Industrial Pollution Control, Inc.
45 Riverside Avenue
Westport, CT 06880
Telephone: 203-227-8497

Established: 1970

Nature and scope of services: "Develop and design waste treatment facilities for industry."

9-c-26 International Acoustical Testing
PO Box 8049
St. Paul, MN 55113
Telephone: 612-633-8434

Established: 1964

Nature and scope of services: Acoustical consulting and testing in field or in our laboratory. OSHA Noise Compliance—fire and furniture testing.

9-d-27 International Hydronics Corp.
Box 910, R-4
Princeton, NJ 08540
Telephone: 201-329-2361

Established: 1966

Nature and scope of services: "Engineering services: Consulting, on-site investigations, process development, engineering design, construction management, and startup supervision—industrial waters and wastes.

"Hyon Waste Management Services, Inc.: Central facility to receive, treat, and ultimately dispose of concentrated industrial wastes."

Additional information: "International Hydronics Corporation specializes in the handling and treatment process for industrial wastes and has been singularly successful in developing both unique and conventional processes for acid mine drainage wastes, ash handling wastes, plating wastes, petrochemical plant wastes, refinery wastes, paper mill wastes, miscellaneous special waste.

"Additional office in Chicago, IL."

9-d-28 Johnson and Anderson, Inc.
Box 1066, 2300 Dixie Highway
Pontiac, MI 48056
Telephone: 313-334-9901

Established: 1946

Nature and scope of services: "Professional engineering services—municipal water, sanitary wastewater, storm water collection, distribution, treatment—industrial water and waste water—Distribution, reuse and treatment."

Additional information: "We have a waste water laboratory and a firm of 200 individuals."

9-d-29 Edward C. Jordan Co., Inc.
379 Congress Street
Portland, ME 04111
Telephone: 207-774-0313

Established: 1873

Nature and scope of services: "Complete engineering, planning and architectural services. Emphasis is on environmental engineering, waste water treatment, water supply, solid waste, impact statements."

Additional information: "Information of a highly technical nature available to other professional organizations."

9-d-30 Raphael Katzen Associates
1050 Delta Avenue
Cincinnati, OH 45208
Telephone: 513-321-8403

Established: 1953

Nature and scope of services: "Technical and economic studies, process design and engineering for chemical process industries including petro-chemicals, organic chemicals, wood chemicals, pulp chemical recovery and by-products, sugar by-products, cryogenics, air pollution, and waste disposal; also consultation on plant operation, improvement, expansion of existing facilities, long-range planning, research and development, site selection and financing, pollution abatement."

9-d-31 Kirkham, Michael and Associates
7300 Woolworth Avenue
Omaha, NE 68124
Telephone: 402-393-5630

Established: 1946

Nature and scope of services: "Complete architectural and engineering design services."

Additional information: "Additional office in Minneapolis, MN."

9-d-32 Kral, Zepf, Freitag and Associates
2830 Victory Parkway
Cincinnati, OH 45206
Telephone: 513-281-7723

Nature and scope of services: "Architectural; mechanical, electrical, structural, and civil engineering; surveying; marketing and planning; airport planning; highway engineering; environmental impact; assessment reports; industrial engineering; construction administration."

Additional information: "If the questions do not require a great deal of research and planning, we will attempt to answer questions concerned with how construction affects the environment. Our most recent publication is an environmental impact study concerned with a nine-mile connector of an interstate. This study includes a corridor report and a study of the air, noise, and water quality.
"Additional office in Crescent Springs, KY."

9-d-33 Langley, McDonald and Overman
484 Newtown Road
Virginia Beach, VA 23462
Telephone: 804-497-8954

Established: 1956

Nature and scope of services: "Surveying and consulting civil engineers; waterfront design, highways, railroads, sewage and water systems."

9-d-34 Laramore, Douglass and Popham
332 South Michigan Avenue
Chicago, IL 60604
Telephone: 312-427-8486

Established: 1937

Nature and scope of services: "Consulting engineering services for utilities and industrial projects."

Additional information: "Additional office in New York City."

9-d-35 Walter C. McCrone Associates, Inc.
2820 South Michigan Avenue
Chicago, IL 60616
Telephone: 312-842-7100

Established: 1956

Nature and scope of services: "(1) Research on methods of characterizing and identifying small particles; microscopical analyses, diffraction, spectroscopy, X-ray fluorescence, mass spectrometry, microprobes, ESCA. (2) Teaching intensive courses in above techniques."

Additional information: "Quarterly newsletter *Insight* available gratis; International journal on applied microscopy *The Microscope* available at $30 per year; four volumes of *The Particle Atlas Two*, a complete text on identification of small particles, available at $240.
"Additional offices in London, England."

9-d-36 Chas. T. Main, Inc.
Southeast Tower, Prudential Center
Boston, MA 02199
Telephone: 617-262-3200

Established: 1893

Nature and scope of services: "Complete design engineering and construction management services for the power industries and general industrial firms together with environmental engineering specialties."

Additional information: "Additional offices in Charlotte, NC; Denver, CO; Portland, OR."

9-d-37 NUS Corp.
4 Research Place
Rockville, MO 20850
Telephone: 301-948-7010

Established: 1961

Nature and scope of services: "Energy and environmental consulting firm."

Additional information: "Publishes quarterly technical newsletter, technical reports, and an annual commercial nuclear-power-plant list."

9-d-38 Orlando Laboratories, Inc.
PO Box 8025-A
Orlando, FL 32806
Telephone: 305-843-1661

Established: 1962

Nature and scope of services: "Water analysis—chemical, physical, and bacteriological. International service. Field and branch laboratory services."

Additional information: "Branch offices in Tampa, FL."

9-d-39 Ostergaard Associates
115 Bloomfield, Avenue
Caldwell, NJ 07006
Telephone: 201-228-0523

Established: 1971

Nature and scope of services: "Consulting services in the field of acoustics—environmental acoustics, industrial noise, building acoustics, transportation noise, hearing conservation."

9-d-40 Pacific Environmental Laboratory
657 Howard Street
San Francisco, CA 94105
Telephone: 415-362-6065

Established: 1920

Nature and scope of services: "Laboratory, analytical, monitoring, and consulting services on water, waste water, air quality, oceanography, oil spill cleanup, corrosion control, and environmental assessment."

Additional information: "Private commercial laboratory. A division of Kennedy Engineers, Inc., with offices in Los Angeles, Tacoma, and Honolulu."

9-d-41 Pavia-Byrne Engineering Corp.
431 Gravier Street
New Orleans, LA 70130
Telephone: 504-581-9451

Established: 1963

Nature and scope of services: "Counseling services for pollution control. Process development, applied research—primarily water pollution control."

9-d-42 Malcom Pirnie, Inc.
2 Corporate Park Drive
White Plains, NY 10602
Telephone: 914-694-2100

Established: 1929

Nature and scope of services: "Consulting environmental engineers, architect environmental impact assessment, pilot studies, design of water and waste water treatment plants, reports and studies."

Additional information: "Offices in Paramus, NJ; Newport News, VA; Columbus, OH; Albany, NY; Waterbury, CT; Hamburg, NY; Washington, DC; Chicago, IL."

9-d-43 Pollution Curbs, Inc.
502 North Prior Avenue
St. Paul, MN 55104
Telephone: 612-647-0151

Established: 1969

Nature and scope of services: "PCI performs ambient and source monitoring services; designs and installs systems and equipment; prepares reports, permit applications and impact statements; all relative to air, water, and/or noise pollution and OSHA situations."

Additional information: "Printed literature about our firm's activities will be sent to interested parties on request."

9-d-44 Pope, Evans, and Robbins, Inc.
11 East 36th Street
New York, NY 10016
Telephone: 212-889-5800

Established: 1906

Nature and scope of services: "Engineering and research related to (1) fluidized bed combustion of coal—a pollution-free method of burning coal—and (2) solid waste management planning."

Additional information: "Additional office in Alexandria, VA."

9-d-45 E. S. Preston Associates, Inc.
939 Goodale Boulevard
Columbus, OH 43212
Telephone: 614-221-7505

Established: 1957

Nature and scope of services: "Consulting engineers with specialties in highway planning and design, plan preparation and construction supervision, structural and community facilities design, environmental evaluation services, and management and financial planning services. Specialists in photogrammetry techniques."

Additional information: "Offices in Charleston, WV and Morristown, TN."

9-d-46 Pyburn & Odom, Inc.
PO Box 267
8178 G.S.R.I. Avenue
Baton Rouge, LA 70821
Telephone: 504-766-6330

Established: 1948

Nature and scope of services: "Pyburn & Odom, Inc., is experienced in the following fields: Hydraulic and hydrologic studies, potamologic studies, laboratory analyses of waste water, sediment and stack gas samples, design of treatment facilities, preparation of environmental assessment reports, design of waterfront structures and submarine pipeline river crossings."

Additional information: "The firm of Pyburn & Odom Consulting Engineers was formed in 1948. It was incorporated under the laws of the State of Louisiana in 1970. Since its inception the firm has provided a wide range of civil engineering services throughout the United States and in several foreign countries, with particular emphasis on water-related projects.

"During the twenty-six years of its existence, our firm has done the engineering work involved in site selection, design and supervision of construction for pipelines and pipeline river crossings, river front structures, stream pollution control and industrial waste disposal, hydrologic and water resources development, and industrial site selection and development. This experience has resulted in the development of precise skills and equipment in each of these activities."

9-d-47 Radiation Management Corp.
Suite 400, Science Center Building #2
3508 Market Street
Philadelphia, PA 19104
Telephone: 215-243-2950

Established: 1969

Nature and scope of services: "A service organization which offers a blend of medical and scientific capabilities to nuclear industry. It covers all aspects of human radiation problems associated with the use of nuclear energy."

Additional information: "Operated in conjunction with the Hospital of the University of Pennsylvania where clinical facilities are located."

9-d-48 Robert and Company Associates
Architects, Engineers, Planners
96 Poplar Street
Atlanta, GA 30303
Telephone: 404-525-8411

Established: 1917

Nature and scope of services: "Full architectual engineering and planning services: architecture and industrial design, city planning, urban design, landscape design, interior design; municipal services, environmental systems, pollution control systems, transportation systems, airports, roads, highway and bridge engineering; structural, civil, mechanical, air-conditioning, and electrical engineering."

Additional information: "Specific areas of expertise are: hospitals, medical facilities of all types; research laboratories; educational facilities, libraries, auditoria, civic and public buildings; industrial, electrical, and textile manufacturing plants; acoustical-environmental reports; airport design; municipal facilities design; mechanical and power studies and facilities. "Additional offices in West Palm Beach, FL."

9-d-49 Rossnagel and Associates
1999 Route 70
Cherry Hill, NJ 08003
Telephone: 609-424-4440

Established: 1970

Nature and scope of services: "Air, water, and noise testing and consulting engineers. Specialists in exhaust, stack and boiler testing; gas analysis testing; water pollution testing; and all phases of consulting in these areas."

Additional information: "Offices in Atlanta, GA."

9-d-50 J. E. Sirrine Company
PO Box 5456, Station B
Greenville, SC 29606
Telephone: 803-271-9350

Established: 1902

Nature and scope of services: "An independent consulting firm that specializes in air, water, and noise pollution abatement. A full range of engineering services needed for environmental studies, process planning, and construction and construction administration are available."

Additional information: "Offices in Raleigh, NC and Houston TX."

9-d-51 Southern Research Institute
2000 Ninth Avenue
South Birmingham, AL 35205
Telephone: 205-323-6592

Established: 1941

Nature and scope of services: "Research in the physical, biological, engineering, and social sciences. Environmental research, especially on control of air and water pollution, is an appreciable part of the Institute's activities."

9-d-52 Spotts, Stevens, and McCoy, Inc.
PO Box 70
Wyomissing, PA 19610
Telephone: 215-376-6581

Established: 1967

Nature and scope of services: "Consulting firm specializing in environmental and municipal engineering, planning, and surveying. Scope of services offered is from initial sampling and evaluation, data collection, and feasibility studies through field surveying, detailed design, and construction management and inspection, to start up and advice on operation of facilities."

Additional information: "Additional offices in Lehighton, PA; Strasburg, PA; Sunbury, PA."

9-d-53 The Stanwick Corp.
3661 East Virginia Beach Boulevard
Norfolk, VA 23502
Telephone: 804-855-8681

Nature and scope of services: "(1) Air pollution study for Staten Island, NY; (2) development of maintenance systems for environmental life systems, Johns Hopkins Hospital, Baltimore, MD; (3) study of water pollution for Van Gool Slaughterhouse, Belgium; (4) study of air pollution in the Belgium plant of the Monroe Manufacturing Company (automobile equipment)."

Additional information: "Our work is international in scope, and an additional office is located in Arlington, VA 22209."

9-d-54 Stearns-Roger, Inc.
Environmental Sciences Division
Box 5888, 700 South Ash
Denver, CO 80217
Telephone: 303-758-1122

Established: Company, 1885; Division, 1970

Nature and scope of services: "We provide the complete scope of environmental services; plant site selection, environmental baseline studies including terrestrial and aquatic ecology studies, socioeconomic studies, meteorological and air quality monitoring, mathematical modeling of single or multiple source emissions, environmental impact analysis, testimony before regulatory bodies and public hearings, etc."

Additional information: "Stearns-Roger, Inc., is an engineering design and construction firm. The Environmental Sciences Division is separate and distinct from the engineering functions of the company. Environmental studies leading to environmental impact statements require the baseline environmental studies as well as engineering studies in parallel, in order that the engineering and environmental alternates be considered simultaneously for optimum plant design."

9-d-55 Trans Urban East Organization, Inc.
444 Central Park West, Suite 20-A
New York, NY 10025
Telephone: 212-663-1213

Established: 1969

Nature and scope of services: "Independent planning and consulting offering comprehensive services in the areas of program planning and development, research, evaluation, administration, management, and training. Four divisions focus on four critical areas: housing and community development, urban health and ecology, economic development, and education."

Additional information: "Offices in Washington, DC."

9-d-56 Charles R. Velzy Associates, Inc.
350 Executive Boulevard
Elmsford, NY 10523
Telephone: 914-592-4750

Established: 1966

Nature and scope of services: "Design and analysis for systems and projects connected with waste water collection and treatment, industrial waste treatment, solid waste collection and disposal, water supply treatment and distribution, drainage and flood control, and air pollution control."

Additional information: "Offices in Mineola and Babylon, NY."

9-d-57 Leonard S. Wegman Co., Inc.
101 Park Avenue
New York, NY 10017
Telephone: 212-686-0500

Established: 1944

Nature and scope of services: "Planning, design, inspection, evaluation, and economic studies; specializations in solid waste management, sewage treatment, water supply, highways and bridges, and waterfront structures."

Additional information: "The firm's principals have testified on national environmental legislation at United States Senate hearings in 1968, 1970, 1972, and 1973."

9-d-58 Wells Laboratories, Inc.
75 North Lewis Avenue
Jersey City, NJ 07306
Telephone: 201-992-3369

Established: 1948

Nature and scope of services: "Life sciences, all phases: microbiology, toxicology, biochemistry, chemistry (analytical research), pathology, physiology, pharmacology, air, water, food, wastes, botanicals, biologicals, disinfectants, cosmetics, etc., etc."

Additional information: "Staff: thirty-five full time, eleven consultants part time; working beauty shop and in-use test facilities on premises; work with advertising agencies; provide testimony; prepare FDA, EPA, departments of Agriculture, Interior, and other government agency requirements. Work with rodents, primates, pisces, dogs, cats, sheep, fowl, and horses. Have a sixty-eight-acre working farm. Prepare unusual sera for research institutions; can undertake basic research in many areas."

9-d-59 Wendel Associates, Consulting Engineers, Planners
7405 Canal Road
Lockport, NY 14094
Telephone: 716-433-5993

Established: 1940

Nature and scope of services: "Studies, plans, reports. Design and construction observation of projects such as master plans, site plans, waste water treatment, water supply, highways, and recreation and residential areas."

9-d-60 Whitman, Requardt and Associates
1304 St. Paul Street
Baltimore, MD 21202
Telephone: 301-727-3450

Established: 1915

Nature and scope of services: "Consulting engineers rendering mechanical, electrical, civil, sanitary, highway, and evaluation engineering services with auxiliary architectural group to industry, the federal government, public utilities, and state and municipal governments."

Additional information: "Complete professional services are available for projects ranging from airports to waste disposal systems."

CHAPTER

10

GOVERNMENT OFFICIALS

———————◆•••◆———————

This listing of government officials contains names, addresses, and telephone numbers of regional Environmental Protection Agency directors and the environmental control officers for each state. The EPA region is noted for each state official. In states without a listing for an environmental control officer, no single individual has been identified by the EPA as being in charge of environmental affairs. This listing contains substantial changes from previously published sources.

EPA REGIONAL OFFICES AND ADMINISTRATORS

10-a-1 Region I
John A. S. McGlennon
John F. Kennedy Federal Building
Room 2203
Boston, MA 02203
Telephone: 617-223-7210

10-a-2 Region II
Gerald M. Hansler
26 Federal Plaza
Room 908
New York, NY 10007
Telephone: 212-264-2525

10-a-3 Region III
Daniel J. Snyder, III
Curtis Building
6th and Walnut Streets
Philadelphia, PA 19106
Telephone: 215-597-9801

10-a-4 Region IV
Jack E. Ravan
1421 Peachtree Street, NE
Atlanta, GA 30309
Telephone: 404-526-5727

10-a-5 Region V
Francis T. Mayo
1 North Wacker Drive
Chicago, IL 60606
Telephone: 312-353-5250

10-a-6 Region VI
Arthur W. Busch
1600 Patterson Street
Suite 1100
Dallas, TX 75201
Telephone: 214-749-1962

10-a-7 Region VII
Jerome H. Svore
1735 Baltimore Avenue
Kansas City, MO 64108
Telephone: 816-374-5493

10-a-8 Region VIII
John A. Green
1860 Lincoln Street
Suite 900
Denver, CO 80203
Telephone: 303-837-3895

10-a-9 Region IX
Paul DeFalco, Jr.
100 California Street
San Francisco, CA 94111
Telephone: 415-556-2320

10-a-10 Region X
Clifford V. Smith
1200 Sixth Avenue
Seattle, WA 98101
Telephone: 206-442-1220

STATE AND TERRITORIAL ENVIRONMENTAL CONTROL OFFICERS

10-b-1 Alabama—EPA Region IV
William T. Willis
Chief Engineer and Director
Bureau of Environmental Health
Alabama Department of Health
State Office Building
Montgomery, AL 36104
Telephone: 205-832-3176

10-b-2 Alaska—EPA Region X
Ernst W. Mueller, Commissioner
Department of Environmental
   Conservation
Pouch O
Juneau, AK 99801
Telephone: 907-586-6721

10-b-3 American Samoa—
   EPA Region IX
Donald Graf, Executive Secretary
Environmental Quality
   Commission
Government of American Samoa
Office of the Governor
Pago Pago, American Samoa 96920

10-b-4 Arizona—EPA Region IX
Dr. James Schumadan,
   Commissioner
Department of Health
1740 West Adams Street
Phoenix, AZ 85007
Telephone: 602-271-5572

10-b-5 Arkansas—EPA Region VI
S. Ladd Davies, Director
Department of Pollution Control
   and Ecology
8001 National Drive
Little Rock, AR 72209
Telephone: 501-371-1701

10-b-6 California—EPA Region
   IX
Claire Dedrick
Secretary for Resources
The Resources Agency
1416 Ninth Street
Sacramento, CA 95814
Telephone: 916-445-5656

10-b-7 Colorado—EPA
Region VIII
Dr. Edward G. Dreyfus,
Executive Director
Department of Health
4210 East 11th Avenue
Denver, CO 80220
Telephone: 303-388-6111 Ext. 315

10-b-8 Connecticut—EPA
Region I
Douglas Costle, Commissioner
Department of Environmental
Protection
State Office Building
165 Capitol Avenue
Hartford, CT 06119
Telephone: 203-566-4030

10-b-9 Delaware—EPA Region III
John Bryson, Secretary
Department of Natural Resources
and Environmental Control
Tatnall Building
Dover, DE 19901
Telephone: 302-678-4403

10-b-10 District of Columbia—
EPA Region III
William C. McKinney, Director
Department of Environmental
Services
412 12th Street, NW
Washington, DC 20004
Telephone: 202-629-3415

10-b-11 Florida—EPA Region IV
Peter P. Baljet, Executive Director
Florida Department of
Pollution Control
2562 Executive Center Circle, East
Montgomery Building
Tallahassee, FL 32301
Telephone: 904-488-4807

10-b-12 Georgia—EPA Region IV
J. Leonard Ledbetter, Director
Environmental Protection Division
Department of Natural Resources
270 Washington Street SW,
Suite 822
Atlanta, GA 30334
Telephone: 404-656-4713

10-b-13 Guam—EPA Region IX
O. V. Natarajan, Ph.D.
Administrator
Environmental Protection Agency
Government of Guam
P.O. Box 2999
Agana, GU 96910
Telephone: 749-9903

10-b-14 Hawaii—EPA Region IX
George Yuen, Director
Department of Health
PO Box 3378
Honolulu, HI 96801
Telephone: 808-548-2211

10-b-15 Idaho—EPA Region X
Dr. James A. Bax, Administrator
Idaho Department of Health
and Welfare
State House
Boise, ID 83701
Telephone: 208-964-2340

10-b-16 Illinois—EPA Region V
Dr. Richard A. Briceland, Director
Illinois Environmental
Protection Agency
2200 Churchill Road
Springfield, IL 62706
Telephone: 217-525-3397

10-b-17 Indiana—EPA Region V
William T. Paynter, M.D.,
    Secretary
State Board of Health
1330 West Michigan Street
Indianapolis, IN 46206
Telephone: 317-633-4000
Oral Hert, Technical Secretary
Stream Pollution Control Board
1330 West Michigan Street
Indianapolis, IN 46206
Telephone: 317-633-4420

10-b-18 Iowa—EPA Region VII
Larry E. Crone, Executive Director
Department of Environmental
    Quality
3920 Delaware, PO Box 3326
Des Moines, IA 50319
Telephone: 515-265-8134

10-b-19 Kansas—EPA Region
    VII
Melville Gray, Director
Division of Environmental Health
Department of Health and
    Environment
Topeka, KS 66620
Telephone: 913-296-3821

10-b-20 Kentucky—EPA
    Region IV
Herman D. Regan, Jr.,
    Commissioner
Bureau of Environmental Quality
Department for Natural Resources
    and Environmental Protection
Capital Plaza Towers
Frankfort, KY 40601
Telephone: 502-564-7030

10-b-21 Louisiana—EPA
    Region VI
None

10-b-22 Maine—EPA Region I
William R. Adams, Director
Department of Environmental
    Protection
State House
Augusta, ME 04330
Telephone: 207-289-2591

10-b-23 Maryland—EPA
    Region III
Herbert Sachs, Director
Department of Natural Resources
State Office Building
Annapolis, MD 21401
Telephone: 301-267-5041

10-b-24 Massachusetts—
    EPA Region I
Dr. Evelyn Murphy, Secretary
Executive Office of
    Environmental Affairs
18 Tremont Street
Boston, MA 02108
Telephone: 617-727-7700

10-b-25 Michigan—EPA Region
    V
A. Gene Gazlay, Director
Department of Natural Resources
Stevens T. Mason Building
Lansing, MI 48926
Telephone: 517-373-1220
Ralph W. Purdy,
    Executive Secretary
Water Resources Commission
Stevens T. Mason Building
Lansing, MI 48926
Telephone: 517-373-3560

10-b-26 Minnesota—EPA
Region V
Grant J. Merritt, Executive
Director
Minnesota Pollution
Control Agency
1935 West County Road B2
Roseville, MN 55113
Telephone: 612-296-5500

10-b-27 Mississippi—EPA
Region IV
Glen Wood, Jr., Executive Director
Mississippi Air and Water
Pollution Control Commission
PO Box 827
Jackson, MI 39205
Telephone: 601-354-6783

10-b-28 Missouri—EPA
Region VII
Kenneth Korch, Director
Division of Environmental Quality
PO Box 1368
Jefferson City, MO 65101
Telephone: 314-751-3241

10-b-29 Montana—EPA
Region VIII
Benjamin Wake, Administrator
Environmental Sciences Division
Department of Health and
Environmental Sciences
Cogswell Building
Helena, MT 59601
Telephone: 406-449-3454

10-b-30 Nebraska—EPA
Region VII
Dan T. Drain, Director
Department of Environmental
Control
1424 P Street
Lincoln, NE 68509
Telephone: 402-471-2186

10-b-31 Nevada—EPA Region IX
Roger Trounday, Director
Department of Health, Welfare
and Rehabilitation
201 South Fall Street
Carson City, NV 89701
Telephone: 702-882-7482

10-b-32 New Hampshire—EPA
Region I
None

10-b-33 New Jersey—EPA
Region II
David J. Bardin, Commissioner
Department of Environmental
Protection
PO Box 1390
Trenton, NJ 08625
Telephone: 609-292-2885

10-b-34 New Mexico—EPA
Region VI
Aaron L. Bond, Director
Environmental Improvement
Agency
PO Box 2348
Santa Fe, NM 87501
Telephone: 505-827-2373

10-b-35 New York—EPA
Region II
Ogden R. Reid, Commissioner
Department of Environmental
Conservation
50 Wolf Road
Albany, NY 12205
Telephone: 518-457-3446

10-b-36 North Carolina—
EPA Region IV
Lewis R. Martin, Director
Division of Environmental
Management

North Carolina Department of
Natural and Economic
Resources
PO Box 27687
Raleigh, NC 27611
Telephone: 919-829-4740

10-b-37 North Dakota—EPA
Region VIII
W. Van Heuvelen, Chief
Environmental Health and
Engineering Services
Department of Health
State Capitol
Bismarck, ND 58501
Telephone: 701-224-2371

10-b-38 Oklahoma—EPA
Region VI
Dr. Thomas D. Peace, Director
Department of Pollution Control
Box 53504
NE 10th and Stonewall
Oklahoma City, OK 73105
Telephone: 405-271-5600

10-b-39 Ohio—EPA Region V
Ned Williams, Director
Ohio Environmental
Protection Agency
361 East Broad Street
PO Box 1049
Columbus, OH 43216
Telephone: 614-469-6126

10-b-40 Oregon—EPA Region X
Kessler Cannon, Director
Department of Environmental
Quality
1234 SW Morrison
Portland, OR 97207
Telephone: 503-229-5696

10-b-41 Pennsylvania—
EPA Region III
Maurice Goddard, Secretary
Department of Environmental
Resources
PO Box 1467
Harrisburg, PA 17120
Telephone: 717-787-2814

10-b-42 Puerto Rico—EPA
Region II
Carlos Jimenez Barber,
Executive Director
Environmental Quality Board
PO Box 11488
Santurce, PR 00910
Telephone: 809-725-5140

10-b-43 Rhode Island—EPA
Region I
None

10-b-44 South Carolina—EPA
Region IV
Dr. E. Kenneth Aycock,
Commissioner
South Carolina Department of
Health and Environmental
Control
2600 Bull Street
Columbia, SC 29201
Telephone: 803-758-5443

10-b-45 South Dakota—EPA
Region VIII
Dr. Allyn O. Lockner, Secretary
Department of Environmental
Protection
State Office Building #2
Pierre, SD 57501
Telephone: 605-224-3351

10-b-46 Tennessee—EPA
   Region IV
J. L. Church, Jr.
Deputy Commissioner for
   Environment
Bureau of Environmental Health
Tennessee Department of
   Public Health
349 Cordell Hull Building
Nashville, TN 37219
Telephone: 615-741-3657

10-b-47 Texas—EPA Region VI
None

10-b-48 Trust Territory of the
   Pacific Islands—EPA Region IX
Sebastian Ongesii, Acting
   Executive Director
Division of Environmental Health
Department of Health Services
Trust Territory of the
   Pacific Islands
Saipan, Mariana Islands 96950

10-b-49 Utah—EPA Region VIII
Lynn M. Thatcher
Department Director of Health and
   Environment Programs
44 Medical Drive
Salt Lake City, UT 84113
Telephone: 801-328-6121
   (FTS 801-524-5500)

10-b-50 Vermont—EPA Region I
Martin Johnson, Secretary
Agency of Environmental
   Conservation
Montpelier, VT 05602
Telephone: 802-828-3357

10-b-51 Virgin Islands—
   EPA Region II
Pedrito François, Director
Division of Natural Resource
   Management
Department of Conservation and
   Cultural Affairs
St. Thomas, VI 00802
Telephone: 809-774-6880

10-b-52 Virginia—EPA Region
   III
None

10-b-53 Washington—EPA
   Region X
John A. Biggs, Director
Department of Ecology
Olympia, WA 98504
Telephone: 206-753-2813

10-b-54 West Virginia—EPA
   Region III
Ira S. Latimer, Jr., Director
Department of Natural Resources
1201 Greenbrier Street
Charleston, WV 25311
Telephone: 304-345-2754

10-b-55 Wisconsin—EPA
   Region V
Lester P. Voight, Secretary
Department of Natural Resources
Madison, WI 53701
Telephone: 608-266-2121

10-b-56 Wyoming—EPA
   Region VIII
Robert E. Sundin, Director
Department of
   Environment Quality
State Office Building
Cheyenne, WY 82002
Telephone: 307-777-7391

ACI Films. 35 West 45th Street, New York, NY 10036

Abelard. Abelard-Schuman, 666 Fifth Avenue, New York, NY 10019

Academic. Academic Press, Inc., 111 Fifth Avenue, New York, NY 10003

Academic Media. Marquis Who's Who, Inc., 200 East Ohio Street, Chicago, IL 60611

Addison-Wesley. Addison-Wesley Publishing Co., Inc., Reading, MA 01867

Air Pollution Control Association. 4400 Fifth Avenue, Pittsburgh, PA 15213

Aldine. Aldine Publishing Company, 529 South Wabash Avenue, Chicago, IL 60605

Allyn & Bacon. Allyn & Bacon, Inc., 470 Atlantic Avenue, Boston, MA 02210

American Association for the Advancement of Science. 1515 Massachusetts Avenue NW, Washington, DC 20005

American Chemical Society. 1155 16th Street NW, Washington, DC 20036

American Conference of Governmental Industrial Hygienists. PO Box 1937, Cincinnati, OH 45202

American Elsevier. American Elsevier Publishing Company Co., Inc., 52 Vanderbilt Avenue, New York, NY 10017

American Enterprise Institute for Public Policy Research. 1150 17th Street NW, Washington, DC 20036

American Forestry Association. 1319 18th Street NW, Washington, DC 20036

American Institute of Chemical Engineers, 345 East 47th Street, New York, NY 10017

American Library Association. 50 East Huron Street, Chicago, IL 60611

American Public Health Association, Inc. 1015 18th Street NW, Washington, DC 20036

American Society for Information Science. 1155 16th Street NW, Washington, DC 20036

American Society for Planning Officials. 1313 East 60th Street, Chicago, IL 60637

Ann Arbor Science. Ann Arbor Science Publishers, Inc., Drawer 1425, Ann Arbor, MI 48106

Annual Reviews. Annual Reviews, Inc., 4139 El Camino Way, Palo Alto, CA 94306

Associated-Sterling Films. 866 Third Avenue, New York, NY 10022

Audio Visual Narrative Arts. PO Box 398, Pleasantville, NY 10570

Ballantine. Ballantine Books, Inc., 201 East 50th Street, New York, NY 10022

Bantam. Bantam Books, Inc., 666 Fifth Avenue, New York, NY 10019

A.S. Barnes. A. S. Barnes & Co., Inc., Forsgate Drive, Cranbury, NJ 08512

Beacon. Beacon Press. 25 Beacon Street, Boston, MA 02108

Benchmark Films. 145 Scarborough Road, Briarcliff Manor, NY 10510

Big Sur Recordings. 117 Mitchell Boulevard, San Rafael, CA 94903

Bobbs-Merrill. The Bobbs-Merrill Co., Inc., 4300 West 62d Street, Indianapolis, IN 46206

Bowker. R. R. Bowker Company, 1180 Avenue of the Americas, New York, NY 10036

Braziller. George Braziller, Inc., One Park Avenue, New York, NY 10016

Brown. William C. Brown Company, Publishers, 2460 Kerper Boulevard, Dubuque, IA 52001

Bureau of National Affairs, Inc. 1231 25th Street NW, Washington, DC 20037

CCM Information. CCM Information Corporation, 866 Third Avenue, New York, NY 10022

CRC Press. 1890 Cranwood Parkway, Cleveland, OH 44128

Caedmon Records. 505 Eighth Avenue, New York, NY 10018

Canadian Broadcasting Corporation. 1500 Bronson Avenue, Ottawa, Ontario K1G 3JG Canada

Canfield. Canfield Press, 850 Montgomery Street, San Francisco, CA 04133

Capricorn. Capricorn Books, 200 Madison Avenue, New York, NY 10016

Catalyst for Environmental Quality. 274 Madison Avenue, New York, NY 10016

Center for Cassette Studies. 8110 Webb Avenue, North Hollywood, CA 91605

Center for the Study of Democratic Institutions. Box 4068, Santa Barbara, CA 93103

Centron Educational Films. 1621 West 9th Street, Lawrence, KS 66044

Childrens Press. 1224 West Van Buren Street, Chicago, IL 60607

Chronicle. Chronicle Books. 870 Market Street, San Francisco, CA 94102

Colorado Associated University Press. University of Colorado, 1424 15th Street, Boulder, CO 80302

Columbia University Press. 562 West 113th Street, New York, NY 10025

Congressional Quarterly, Inc. 1414 22d Street NW, Washington, DC 20037

The Conservation Foundation. 1717 Massachusetts Avenue NW, Washington, DC 10036

Contemporary/McGraw Hill Films. 1221 Avenue of the Americas, New York, NY 10020

Cornell University Press. 124 Roberts Place, Ithaca, NY 14850

Council of Planning Librarians. PO Box 229, Monticello, IL 61856

Coward, McCann. Coward, Mc-Cann & Geoghegan, Inc., 200 Madison Avenue, New York, NY 10016

Cowles. Cowles Book Co., Inc., 114 West Illinois Avenue, Chicago, IL 60610

T. Y. Crowell. Thomas Y. Crowell Company, Inc., 666 Fifth Avenue, New York, NY 10019

Crowell-Collier. Macmillan, Inc., 866 Third Avenue, New York, NY 10022

Crown. Crown Publishers, Inc., 419 Park Avenue South, New York, NY 10016

Damon Corporation. 80 Wilson Way, Westwood, MA 02090

Day. The John Day Company, 666 Fifth Avenue, New York, NY 10019

Defenders of Wildlife. 2000 N Street NW, Washington, DC 20036

Delacorte. Delacorte Press, One Dag Hammarskjold Plaza (245 East 47th Street), New York, NY 10017

Dell. Dell Publishing Co., Inc., One Dag Hammarskjold Plaza (245 East 47th Street), New York, NY 10017

Denoyer-Geppert Audio Visuals. 5235 Ravenswood Avenue, Chicago, IL 60640

Dial. The Dial Press, One Dag Hammarskjold Plaza (245 East 47th Street), New York, NY 10017

Dickenson. Dickenson Pub. Co., Inc., 16561 Ventura Boulevard, Suite 215G, Encino, CA 91316

Dodd, Mead. Dodd, Mead & Company, 79 Madison Avenue, New York, NY 10016

Doubleday. Doubleday & Company, Inc., 245 Park Avenue, New York, NY 10017

Dryden. Dryden Press, 901 North Elm, Hinsdale, IL 60521

Dutton. E. P. Dutton & Co., Inc., 201 Park Avenue South, New York, NY 10003

Duxbury. Duxbury Press, 6 Bound Brook Court, North Scituate, MA 02060

Earth Island. 20 New Cavendish Street, London W1, England

Ecologist. 75 Molesworth Street, Wadebridge, Cornwall PL27 7DS England

Educational Resources Division, Educational Design. 47 West 13th Street, New York, NY 10011

Encyclopedia Britannica Educational Corporation. 425 North Michigan Avenue, Chicago, IL 60611

Environment Information Center, Inc. 124 East 39th Street, New York, NY 10016

Environmental Action, Inc. Room 731, 1346 Connecticut Avenue NW, Washington, DC 20036

The Environmental Law Center. Boston College Law School, Brighton, MA 02135

Environmental Law Institute. Suite 614, 1346 Connecticut Avenue NW, Washington, DC 20036

Environmental Science Service Division, E.R.A., Inc. Stamford, CT

Environmental Studies Institute, International Academy at Santa Barbara. Riverside Campus, 2048 Alameda Padre Serra, Santa Barbara, CA 93103

Europa. Europa Publications, Ltd., 18 Bedford Square, London WC1B 3JN England

Evans. M. Evans & Co., Inc., 216 East 49th Street, New York, NY 10017

Federation of American Societies for Experimental Biology. 9650 Rockville Pike, Bethesda, MD 20014

Film Fair Communications. 10900 Ventura Boulevard, Studio City, CA 91604

Follett. Follett Publishing Company, 1010 West Washington Boulevard, Chicago, IL 60607

Fontana. 14 St. James's Place, London, SW 1, England

Four Winds. Four Winds Press, Scholastic Magazines, Inc., 50 West 44th Street, New York, NY 10036; order department: 906 Sylvan Avenue, Englewood Cliffs, NJ 07632

Free Press. 866 Third Avenue, New York, NY 10022

Freeman. W. H. Freeman and Company, Publishers, 660 Market Street, San Francisco, CA 94104

Friends of the Earth. Friends of the Earth, Inc., 529 Commercial Street, San Francisco, CA 94111

Glencoe. Glencoe Press, Benziger Bruce & Glencoe, Inc., 8701 Wilshire Boulevard, Beverly Hills, CA 90211

Golden Gate. Golden Gate Junior Books, 9622 Hollywood Boulevard, Los Angeles, CA 90028

Golem. The Golem Press, PO Box 1342, Boulder, CO 80302

Goodyear. Goodyear Publishing Co. Inc., 15113 Sunset Boulevard, Pacific Palisades, CA 90272

Government Printing Office. U.S. Government Printing Office, Division of Public Documents, Washington, DC 20402

Graphics Management. Graphics Management Corporation, 1101 16th Street NW, Washington, DC 20036

Green Eagle. Green Eagle Press, 99 Nassau Street, New York, NY 10038

Greene. Stephen Greene Press, Box 1000, Brattleboro, VT 05301

Grosset. Grosset & Dunlap, Inc., 51 Madison Avenue, New York, NY 10010

Grossman. Grossman Publishers, 625 Madison Avenue, New York, NY 10022

Hafner. Hafner Press, 866 Third Avenue, New York, NY 10022

Harcourt. Harcourt Brace Jovanovich, Inc., 757 Third Avenue, New York, NY 10017

Harper. Harper & Row, Publishers, 10 East 53rd Street, New York, NY 10022

Hart. Hart Publishing Co., Inc., 15 West Fourth Street, New York, NY 10012

Harvest. Harvest House, Ltd., Publishers, 4795 St. Catherine Street West, Montreal, P.Q. G3Z 2B9 Canada

Hawkhill Associates. Black Earth, WI 53515

Hawthorn. Hawthorn Books, Inc., 260 Madison Avenue, New York, NY 10016

Heath. D. C. Heath & Company, 125 Spring Street, Lexington, MA 02173

Heldref Publications. 4000 Albemarle Street NW, Washington, DC 20016

Her Majesty's Stationery Office. U.S. agent: Pendragon House, Inc., 220 University Avenue, Palo Alto, CA 94301

Alfred Higgins Productions. 9100 Sunset Boulevard, Los Angeles, CA 90069

Holt. Holt, Rinehart and Winston, Inc., 383 Madison Avenue, New York, NY 10017; films: Media Department

Houghton Mifflin. Houghton Mifflin Company, 1 Beacon Street, Boston, MA 02107

Indiana University Audio Visual Center. Bloomington, IN 47401

Indiana University Press. 10th and Morton Streets, Bloomington, Ind. 47401

International Film Bureau. 332 South Michigan Avenue, Chicago, IL 60604

International Publishers. International Publishers Co., Inc., 381 Park Avenue South, New York, NY 10016

Intext. Intext Educational Publishers. 666 Fifth Avenue, New York, NY 10019

Johns Hopkins University Press. Baltimore, MD 21218

Jossey-Bass. Jossey-Bass, Inc., Publishers, 615 Montgomery Street, San Francisco, CA 94111

Knight. Charles Knight & Co., 11-12 Bury Street, London EC3A 5AP England

Knopf. Alfred A. Knopf, Inc., 201 East 50th Street, New York, NY 10022

Knox. John Knox Press, 341 Ponce de Leon Avenue NE, Atlanta, GA 30308

Lake Erie Environmental Studies. State University College, Fredonia, NY 14063

Learning Corporation of America. 711 Fifth Avenue, New York, NY 10022

Lexington Books. D. C. Heath & Company, 125 Spring Street, Lexington, MA 02173

Lippincott. J. B. Lippincott Company, East Washington Square, Philadelphia, PA 19105

Little, Brown. Little, Brown and Company, 34 Beacon Street, Boston, MA 02106

Lothrop. Lothrop, Lee & Shepard Company, 105 Madison Avenue, New York, NY 10016

MIT Press. 28 Carleton Street, Cambridge, MA 02142

McGraw-Hill. McGraw-Hill Book & Education Services Group, 1221 Avenue of the Americas, New York, NY 10020

McKay. David McKay Co., Inc., 750 Third Avenue, New York, NY 10017

Macmillan. Macmillan Publishing Co., Inc., 866 Third Avenue, New York, NY 10022

Macmillan (London). Macmillan & Co., Ltd., Little Essex Street, London WC2R 3LF England

Macmillan Films. 34 MacQuesten Parkway, Mount Vernon, NY 10550

Mason & Lipscomb. Mason/ Charter Publishers, Inc., 384 Fifth Avenue, New York, NY 10018

Mass Media Associates. 2116 North Charles Street, Baltimore, MD 21218

Merrill. Charles E. Merrill Publishing Company, 1300 Alum Creek Drive, Columbus, OH 43216

Messner. Julian Messner, 1 West 39th Street, New York, NY 10018

Methuen. Methuen & Co., Ltd., 11 New Fetter Lane, London EC4P 4EE England

Morrow. William Morrow & Co., Inc., 105 Madison Avenue, New York, NY 10016

Mosby. The C. V. Mosby Company, 11830 Westline Industrial Drive, St. Louis, MO 63141

Mountain Press. Mountain Press Publishing Co., 279 West Front Street, Missoula, MT 59801

NBC Educational Enterprises. 30 Rockefeller Plaza, New York, NY 10020

Nash. Nash Publishing Corporation, 1 DuPont Street, Plainview, NY 11803

National Academy of Sciences. 2101 Constitution Avenue, Washington, DC 20418

National Audubon Society. 950 Third Avenue, New York, NY 10022

National Film Board of Canada. U.S. office: 1251 Avenue of the Americas, New York, NY 10020; in Canada, apply to nearest local office.

National Foundation for Environmental Control, Inc. 152 Tremont Street, Boston, MA 02111

National Geographic Society. 17th and M Streets NW, Washington, DC 20036; for films, Department 76.

National Information Center for Educational Media (NICEM). University of Southern California, University Park, Los Angeles, CA 90007

National Parks Association. 1701 18th Street NW, Washington, DC 20009

National Press Books. Mayfield Publishing Company, 285 Hamilton Avenue, Palo Alto, CA 94301

National Science Foundation. 1800 G Street NW, Washington, DC 20550

National Science Teachers Association. Publications Department, 1201 16th Street NW, Washington, DC 20036

National Technical Information Service, U. S. Department of Commerce. Springfield, VA 22151

National Textbook. National Textbook Company, 8259 Niles Center Road, Skokie, IL 60076

National Wildlife Federation, Inc. 1412 16th Street NW, Washington, DC 20036

Natural History. Natural History Press, 501 Franklin Avenue, Garden City, NY 11530

Nelson. Thomas Nelson & Sons, Ltd., Publishers, 36 Park Street, London W1Y 4DE England; Thomas Nelson Inc., 407 Seventh Avenue, Nashville, TN 37203

Nelson-Hall. Nelson-Hall Company, 325 West Jackson Boulevard, Chicago, IL 60606

New York Graphic Society. New York Graphic Society Ltd., 11 Beacon Street, Boston, MA 02108

New York Times, Educational Division. 229 West 43d Street, New York, NY 10036

North American. North American Publishing Co., 134 North 13th Street, Philadelphia, PA 19107

Norton. W. W. Norton & Company, Inc., 500 Fifth Avenue, New York, NY 10036

Noyes. Noyes Data Corporation, Mill Road and Grand Avenue, Park Ridge, NJ 07656

Oceana. Oceana Publications, Inc., Dobbs Ferry, NY 10522

Office of Community Outreach, University of Wisconsin—Green Bay. Green Bay, WI 54302

Oregon State University Press. Box 689, Corvallis, OR 97330

Output Systems. Output Systems Corporation, 2300 South 9th Street, Arlington, VA 22204

Oxford. Oxford University Press, Inc., 200 Madison Avenue, New York, NY 10016; Ely House, 37 Dover Street, London W1X 4AH England

Pantheon. Pantheon Books, Inc., 201 East 50th Street, New York, NY 10022

Parents'. Parents' Magazine Press. 52 Vanderbilt Avenue, New York, NY 10017

Parker Brothers. 190 Bridge Street, Salem, MA 01970

Parker Publishing. Parker Publishing Co., Englewood Cliffs, NJ 07632; mailing address: West Nyack, NY 10994

Pegasus. The Bobbs-Merrill Co., Inc., 4300 West 62d Street, Indianapolis, IN 46268

Penguin. Penguin Books Inc., 7110 Ambassador Road, Baltimore, MD 21207

Perennial Education. 1825 Willow Road, Northfield, IL 60093

Playboy Press. 919 North Michigan Avenue, Chicago, IL 60611

Plenum. Plenum Publishing Corporation, 227 West 17th Street, New York, NY 10011

Pocket. Pocket Books, 630 Fifth Avenue, New York, NY 10020

Pollution Abstracts. 7611 Convoy Court, San Diego, CA 92037

Population Commission Films. Fisher Film Group, 216 East 49th Street, New York, NY 10017

Praeger. Praeger Publishers, Inc., 111 Fourth Avenue, New York, NY 10003

Prentice-Hall. Prentice Hall, Inc., Englewood Cliffs, NJ 07632

Princeton University Press. Princeton, NJ 08540

Public Administration Services. 1313 East 60th Street, Chicago, IL 60637

Public Affairs Information Service, Inc., 11 West 40th Street, New York, NY 10018

Putnam. G. P. Putnam's Sons, 200 Madison Avenue, New York, NY 10016

QED Productions. 2921 West Alameda Avenue, Burbank, CA 91505

Quadrangle. Quadrangle/The New York Times Book Co., 10 East 53d Street, New York, NY 10022

Rand McNally. Rand McNally & Company, Box 7600, Chicago, IL 60680

Random House. Random House, Inc., 201 East 50th Street, New York, NY 10022

Regnery. Henry Regnery Company, 180 North Michigan Avenue, Chicago, IL 60601

Reilly & Lee. Reilly & Lee Books, Henry Regnery Company, 180 North Michigan Avenue, Chicago, IL 60601

Reinhold. Van Nostrand Reinhold Company, 450 West 33d Street, New York, NY 10001

Rodale. Rodale Press, Book Division, 33 East Minor Street, Emmaus, PA 18049

Ronald. The Ronald Press Company, 79 Madison Avenue, New York, NY 10016

Routledge & Kegan Paul. Routledge & Kegan Paul, Ltd., 68-74 Carter Lane, London EC4V 5EL England; orders to Broadway House, Reading Road, Henley-on-Thames, Oxon, England

Russell & Russell. Russell & Russell, Publishers, 122 East 42d Street, New York, NY 10017

Sage Hill. Sage Hill Publishers, Inc., 116 Washington Avenue, Albany, NY 12210

St. Martin's. St. Martin's Press, Inc., 175 Fifth Avenue, New York, NY 10010

Sargent. Porter E. Sargent, Inc., Publishers, 11 Beacon Street, Boston, MA 02108

Saunders. W. B. Saunders Company, West Washington Square, Philadelphia, PA 19105

Schenkman. Schenkman Publishing Co., Inc., 3 Mount Auburn Place, Harvard Square, Cambridge, MA 02138

Schocken. Schocken Books Inc., 200 Madison Avenue, New York, NY 10016

Scientists' Institute for Public Information. 438 North Skinker Boulevard, St. Louis, MO 63130

Scott, Foresman. Scott, Foresman and Company, 1900 East Lake Avenue, Glenview, IL 60025

Scribner. Charles Scribner's Sons, 597 Fifth Avenue, New York, NY 10017

Seabury. The Seabury Press, Inc., 815 Second Avenue, New York, NY 10017

Sentinel. Sentinel Books, Publishers, Inc., 17 East 22d Street, New York, NY 10010

Serina. Serina Press, 70 Kennedy Street, Alexandria, VA 22305

Sierra Club. Sierra Club Books, 1050 Mills Tower, San Francisco, CA 94104

Simon & Schuster. Simon & Schuster, Inc., 630 Fifth Avenue, New York, NY 10020

Sinauer. Sinauer Associates, Inc., 20 Second Street, Stamford, CT 06905

Sloane. William Sloane Associates, current address unknown.

Smithsonian Institution Press. Washington, DC 20560

Society for Visual Education. 1345 Diversey Parkway, Chicago, IL 60614

Southern Baptist Convention, Radio and TV Committee. 511 River Street, Missoula, MT 54801

Special Reports, Inc. Current address unknown.

Springer. Springer-Verlag New York, Inc., 175 Fifth Avenue, New York, NY 10010

Stackpole. Stackpole Books, Cameron and Kelker Streets, Harrisburg, PA 17105

Steck-Vaughn. Steck-Vaughn Company, Box 2028, Austin, TX 78767

Stein & Day. Stein & Day Publishers, Scarborough House, Briarcliff Manor, NY 10510

Syracuse University Libraries. Syracuse, NY 13210

Texture Films, Inc. 1600 Broadway, New York, NY 10019

Thomas. Charles C. Thomas, Publisher, 301-27 East Lawrence Avenue, Springfield, IL 62717

Time-Life Books. Time & Life Building, Rockefeller Center, New York, NY 10020

Time/Life Education. 43 West 16th Street, New York, NY 10011

UNESCO. Place de Fontenoy, F-75700 Paris, France

Unipub. Box 433, Murray Hill Station, New York, NY 10016

Universe. Universe Books, 381 Park Avenue South, New York, NY 10016

Universitetsforlaget. PO Box 142, Boston, MA 02113

University Microfilms. 300 North Zeeb Road, Ann Arbor, MI 48106

University of Alaska Press. University Campus, College, AK 99735

University of California Press. 2223 Fulton Street, Berkeley, CA 94720

University of Chicago Press. 5801 Ellis Avenue, Chicago, IL 60637

University of Georgia Press. Athens, GA 30602

University of Illinois Graduate School of Library Science. Urbana, IL 61801

University of Illinois Press. Urbana, IL 61801

University of Miami Press. Drawer 9088, Coral Gables, FL 33124

University of Michigan Press. Ann Arbor, MI 48106

University of Oklahoma Press. 1005 Asp Avenue, Norman, OK 73069

University of Pennsylvania Press. 3933 Walnut Street, Philadelphia, PA 19174

University of Toronto Press. St. George Campus, Toronto, Ontario M5S 1A6 Canada

University of Utah Press. Building 513, Salt Lake City, UT 84112

University of Washington Press. 1405 NE 41st Street, Seattle, WA 98195

University of Wisconsin Press. Box 1379, Madison, WI 53701

Van Nostrand Reinhold. Van Nostrand Reinhold Company, 450 West 33d Street, New York, NY 10001

Vanderbilt University Press. Nashville, TN 37235

Viking. The Viking Press, Inc., 625 Madison Avenue, New York, NY 10022

Vocational Guidance. Vocational Guidance Manuals, Inc., 620 South Fifth Street, Louisville, KY 40202

Wadsworth. Wadsworth Publishing Co., Inc., Belmont, CA 94002

Walker. Walker & Company, 720 Fifth Avenue, New York, NY 10019

Ward Ritchie. The Ward Ritchie Press, 474 South Arroyo Parkway, Pasadena, CA 91105

Washington Square. Washington Square Press, 630 Fifth Avenue, New York, NY 10020

Water Information Center. 44 Sintsink Drive East, Port Washington, NY 11050

Watts. Franklin Watts, Inc., 730 Fifth Avenue, New York, NY 10019

Westminster. The Westminster Press, Witherspoon Building, Philadelphia, PA 19107

The Wilderness Society. 1901 Pennsylvania Avenue NW, Washington, DC 20006

Wiley. John Wiley & Sons, Inc., 605 Third Avenue, New York, NY 10016

Wiley-Interscience. John Wiley & Sons, Inc., 605 Third Avenue, New York, NY 10016

Williams & Wilkins. The Williams & Wilkins Company, 428 East Preston Street, Baltimore, MD 21202

Wilson. The H. W. Wilson Company, 950 University Avenue, Bronx, NY 10452

Winthrop. Winthrop Publishers, Inc., 17 Dunster Street, Cambridge, MA 02138

Woodrow Wilson International Center for Scholars. Smithsonian Institution, Washington, DC 20560

World. The World Publishing Company, 110 East 59th Street, New York, NY 10022

Yale University Press. 302 Temple Street, New Haven, CT 06511

York. York Press, Inc., 6803 York Road, Baltimore, MD 21212

Young Scott. Young Scott Books, Addison-Wesley Publishing Co., Inc., Reading, MA 01867

INDEX

Materials are indexed by number rather than by page. The initial digit, preceding the hyphen, indicates the nature of the material. Thus, the initial number 1 indicates an action guide; 2, an index; 3, a reference book; 4, a history; 5, a monograph; 6, a government publication; 7, a nonprint item; 8, a periodical; 9, an organization; and 10, a government official.

Numbers in brackets refer to the pages of the indexed item, not to the pages of this book.